PARTICIPATORY GEOSPATIAL DEVELOPMENT USING PYTHON

RAVISH BAPNA

M. Tech. (Geoinformatics)

Indian Institute of Technology Kanpur, India

First published: September 2012

In memory of my beloved uncle
Late Shri Devendra Kumar Bafna

PREFACE

This book has been written with an aim to promote open source computing in Geoinformatics domain. It deals with geospatial data processing using open source Python binding of various libraries, with focus on GDAL/OGR package. One can find lot of scattered information on internet in this regard, but I felt lack of systematic reading material which a newbie to open source geospatial data computing must have under his belt. The book is written with an assumption that the reader has basic programming skills in Python.

I would like to thank my family for keeping me motivated in one or the other way. It would be unfair, if I miss to acknowledge my mentors, Dr. Onkar Dikshit (Professor, Indian Institute of Technology Kanpur, India) and Dr. Bharat Lohani (Associate Professor, Indian Institute of Technology Kanpur, India) for introducing me to Geoinformatics discipline.

Author
Ravish Bapna

TABLE OF CONTENTS

LIST OF ILLUSTRATIONS

LIST OF TABLES

Chapter 1
INTRODUCTION

What is the first thought that sweep across the mind of a GIS Analyst when he or she comes across the term "Geospatial data processing"? The person starts thinking of geospatial data, software, and computer; and when it is about choosing software to process data, one has a strong inclination towards leading commercial softwares like ArcGIS (product of Environmental Systems Research Institute, Inc.), ERDAS (product of Intergraph Corporation), ENVI (product of Exelis Visual Information Solution) etc. This book provides a walk-through on how a GIS Analyst can process (read, write, visualize and derive results) geospatial data using open source libraries having Python language binding. It is being assumed that the reader of this book has basic understanding of Python language.

1.1. OPEN SOURCE SOFTWARE

Before stepping into the world of open source scripting modules, one should try to understand the definition of open source software given by *Open Source Initiative* (OSI). It is a non-profit corporation with global scope, formed to educate about and advocate the benefits of open source software, and to build bridges among different constituencies in the open source community.

Open source software is defined as software whose source code is made available under a license that allows the modification, and re-distribution of the software at will. The precise definition of open source software can be found at OSI website. Sometimes a distinction is made between open source software and free software.

1.2. OPEN GEOSPATIAL CONSORTIUM

Open Geospatial Consortium (OGC) is a consortium of over 390 companies, NGOs, research organizations, agencies and universities with a mission to serve as a global forum for the collaboration of developers and users of spatial data products and services, and to advance the development of international standards for geospatial interoperability.

It is important not to confuse open source with open standards. They are entirely different. The special license that govern use and sale of open source software exist not to ensure profits to the software's owner, but to ensure that the software's source code remains in the public domain, though companies are allowed to sell products that include some or all of the source code. Open source software is usually developed not by a single company but by a distributed, informal team of developers. Open source software developers use OGC standards for the same reasons commercial developers use them, to make their products interoperate with others.

OGC provides a community consensus process to solve the difficult interoperability issues in the geospatial marketplace. Some users need to share and reuse geospatial content in order to decrease cost, get more or better information, and increase the value of data holding, which can be addressed by cooperation among technology users and providers. OGC brings together geoprocessing technology users and vendors, and provides a formal structure for achieving consensus on the specifications.

1.3. PYTHON

Python is an open source (freely usable and distributable, even for commercial use) programming language that can be used for developing wide variety of applications. Python is available for all major operating systems: Windows, Linux/Unix, Mac, etc. Python can also be embedded as a scripting language in many applications. For example, nowadays, ArcGIS, which is well known commercial GIS software, supports Python scripting. The scripts discussed in this book are written and tested using Python 2.7.2.

Python was created in the early 1990s by *Guido van Rossum* at Stichting Mathematisch Centrum (CWI, see *http://www.cwi.nl/*) in the Netherlands as a successor of a language called ABC. In 2001, the *Python Software Foundation* (PSF, see *http://www.python.org/psf/*) was formed; a non-profit organization created specifically to own Python related intellectual property. Python is also said to be named after the British comedy group *Monty Python*. Python has the right combination of performance and features that make writing programs both fun and easy. Some of the features of Python are listed below:

- Python is a simple and easy to learn.
- Python implementation is under an open source license that makes it freely usable and distributable, even for commercial use.
- Python works on many platforms such as Windows, Linux, etc.
- It is an interpreted language.
- It is an object-oriented language.
- Extensions and modules easily written in C, C++ (or Java using Jython, or .NET language using IronPython).
- Embeddable within applications as a scripting interface.
- Python has a comprehensive set of packages to accomplish various tasks.

1.4. GDAL/OGR

GDAL (pronounced as *gee-doll*) stands for *Geospatial Data Abstraction Library*, is a library intended for working with geospatial raster data, while there is an OGR library for processing vector data. Theoretically, these two are separate packages, but currently they reside in same source tree of GDAL (as shown in figure 1-1). So one cannot download and install GDAL and OGR independently. To avert confusion, the combined library will be referred as GDAL/OGR. This project was started in late 1998 by *Frank Warmerdam*, and he worked as an independent professional on this library.

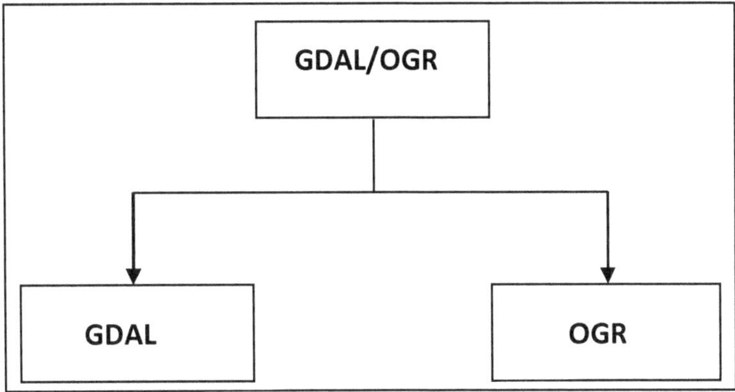

Figure 1-1: Components of GDAL/OGR library

GDAL/OGR is written in C++, but it can also be accessed from Python, Java, C#, VB6 etc. GDAL supports more than 50 raster formats, and OGR supports over 20 vector formats. The library also has many command-line utilities for data translation, image warping, image subsetting, and various other common tasks. It runs on modern versions of Unix, Linux, Solaris, Mac OS X, and most versions of Microsoft Windows, supporting both 32-bit and 64-bit architectures. The library is used by many applications, such as ArcGIS, GRASS, QGIS, and ILWIS etc.

GDAL/OGR library has been maintained by OSGeo, which stands for *Open Source Geospatial Foundation*. OSGeo is an independent non-profit legal entity established to support the needs of the open source geospatial community, which serves as an organizing body, a public technology commons, a development community manager, and event sponsor. The foundation respects the important role that proprietary software plays in geospatial industry, and is not trying to get rid of it, nor the companies that produce it. However, the foundation takes a stance that free and open source software can and should play an important role in the geospatial industry. Furthermore, having quality open source alternatives to proprietary software can be good for the end user, the industry, and even the proprietary software vendors. In fact, some proprietary geospatial softwares are built on open source softwares to some extent.

GDAL/OGR library is distributed under the terms of X11/ MIT style open source license by the *Open Source Geospatial Foundation*. It gives permission to everyone to download, modify, redistribute, build proprietary commercial software using GDAL/OGR source code; no permission from *Frank Warmerdam*, OSGeo Foundation or anyone else is required. But some portion is under slightly different licensing terms, for instance, licensing terms slightly vary for libpng, libjpeg, libtiff, libgeotiff etc. Some external libraries which can be optionally used by GDAL/OGR are under radically different licenses.

1.5. INSTALLATION IN WINDOWS

There are many different ways to install Python and various packages/toolkits; the best approach depends upon the operating system one is using, what is already installed, and how the person intends to use it. To avoid wading through all the details, the easiest approach is to use one of the pre-packaged Python distributions that provide built-in required libraries. An excellent choice for Windows operating system user is Python(x,y), which bundles GDAL/OGR, Matplotlib (plotting library) and lots of other useful tools. The executable file can be downloaded and then installed from the link: *http://code.google.com/p/pythonxy/*. The scripts discussed in this book are tested using version 2.7.2.3 of Python(x,y).

1.6. PYTHON(X,Y)

Python(x,y) is a free scientific and engineering development software for numerical computations, data analysis and data visualization based on Python programming language and Spyder interactive development environment, the launcher (current version 2.7.2.3) is shown in figure 1-2. The main features of Python(x,y) are:

- Bundled with scientific oriented Python libraries and development environment tools.
- Extensive documentation of various Python packages.
- Providing an all-in-one setup program, so that the user can install or uninstall all these packages and features by clicking one button only.

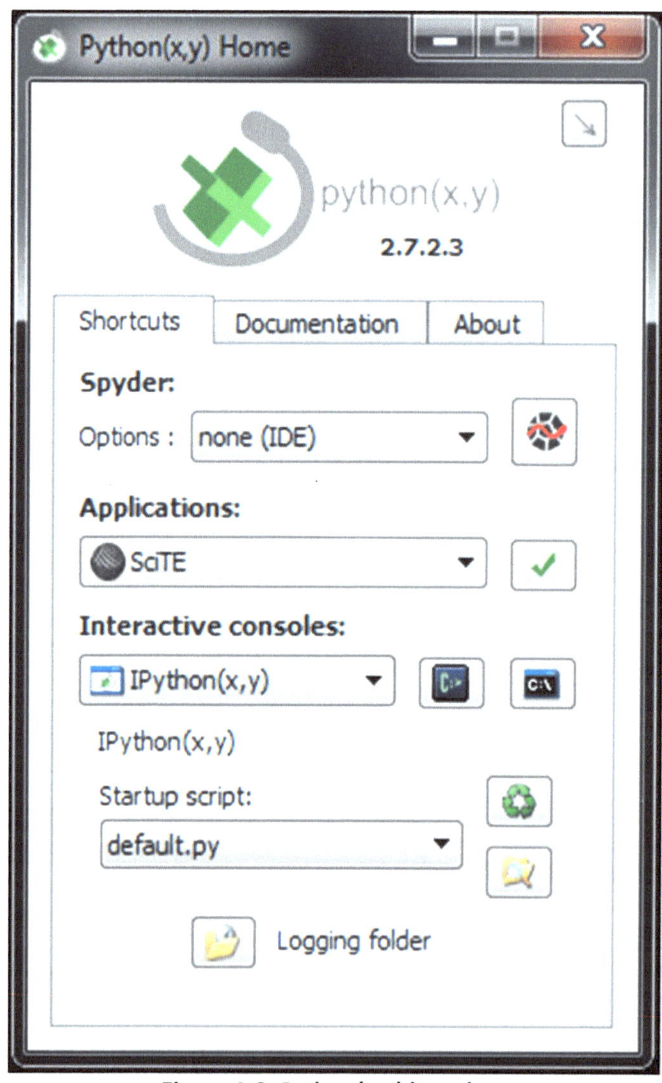

Figure 1-2: Python(x,y) launcher

1.7. SPYDER

Spyder (previously known as *Pydee*) stands for *Scientific PYthon Development EnviRonment* (shown in figure 1-3), and it is a powerful interactive development environment for the Python language with advanced editing, interactive testing, debugging and introspection features. It is also a numerical computing environment, having support of IPython (enhanced interactive Python interpreter) and popular Python libraries such as NumPy, Matplotlib (interactive 2D/3D plotting) etc. Some of the key features are:

- Syntax colouring (or highlighting).
- Typing helpers like automatically inserting closing parentheses etc.
- Contains IPython interpreter.
- Contains basic terminal command window.

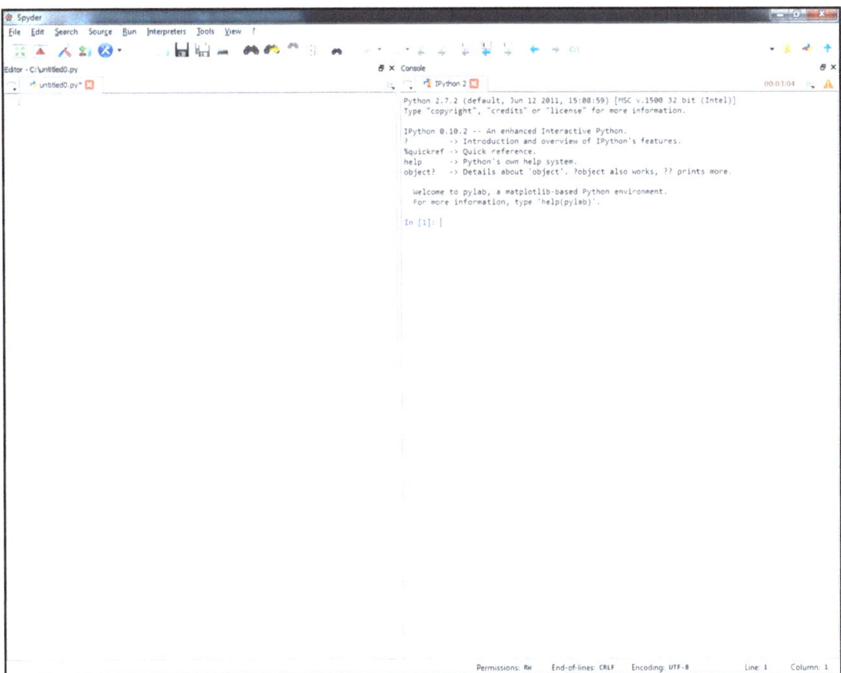

Figure 1-3: Spyder development environment

Spyder runs on all major platforms, and the easiest way to install Spyder in:

- Windows: using an executable installer (*http://spyderlib.googlecode.com*) or through Python(x,y) (*http://www.pythonxy.com*).
- Mac OS X.
- GNU/Linux: through package manager.

1.8. IPYTHON

One of Python's useful feature is its interactive interpreter. This system allows fast testing of ideas without the overhead of creating test files as is typical in most programming languages. However, the interpreter supplied with the standard Python distribution is somewhat limited for extended interactive use. IPython (also packaged in Python(x,y)) provides a rich toolkit which helps in making the most out of Python, with:

- Enhanced Python shell for interactive computing.
- Support for interactive data visualization and use of Graphical User Interface (GUI) toolkits.
- Embeddable interpreter to load into your own projects.
- Easy to use, high performance tools for parallel computing.

1.9. NUMPY

NumPy is a Python library that provides multi-dimensional array object, various derived objects (such as masked arrays), and an assortment of routines for fast operations on arrays, including mathematical, logical, shape manipulation, sorting, selecting, basic linear algebra, basic statistical operations, random simulation and much more. NumPy is packaged in the Python(x,y) setup, so one does not need to install it separately. There are several important differences between NumPy arrays and the standard Python sequences:

- NumPy arrays have a fixed size at creation, unlike Python lists (which can grow dynamically).
- All elements in a NumPy array are required to be of the same data type, and thus, will be of the same size in memory. There is an exception where one can have arrays of (Python, including NumPy) objects, thereby allowing for arrays of different sized elements.
- NumPy arrays facilitate advanced mathematical and other types of operations on large numbers of data. Typically, such operations are executed more efficiently and with less code than Python's built-in sequences.

1.10. OSGEO-LIVE

OSGeo-Live is a self-contained bootable DVD, USB thumb drive or virtual machine based on Xubuntu (operating system), which allows an individual to try a wide variety of open source geospatial software without the compulsion of installing anything. It is composed entirely of free software, sample datasets and documentation which can be freely distributed, duplicated and passed around. One can download OSGeo-Live from the link: *http://live.osgeo.org/en/download.html*. To try out the applications, simply:

i. Insert DVD or USB thumb drive in computer or virtual machine.
ii. Reboot computer (verify boot device order, if necessary).
iii. Press Enter to startup and login.
iv. Try applications from the *Geospatial* menu (as shown in figure 1-4).

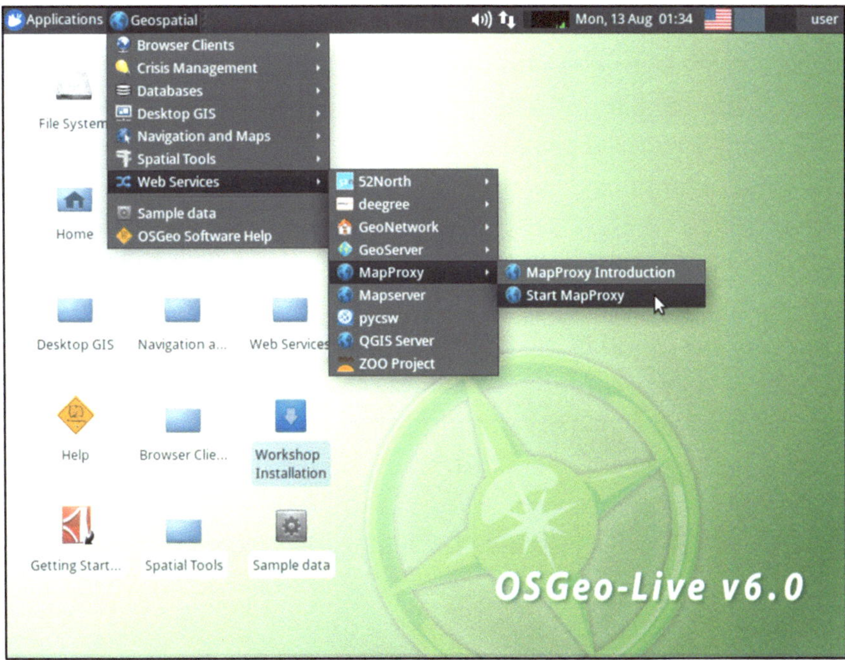

Figure 1-4: Desktop of OSGeo-Live (version 6.0)

1.11. REST OF THE BOOK

Chapter 1 gives an overview of open source packages, namely, GDAL/OGR, NumPy, interactive development environment *Spyder*; interactive toolkit *IPython*, and information on Python(x,y), which has pre-packed Python binding of various libraries.

Chapter 2 enumerates various approaches of representing spatial reference system, namely, EPSG dataset and WKT format. It also narrates about PROJ and Pyproj packages, which are used for cartographic transformations and geodetic computations.

Chapter 3 discusses about colours and map projections supported by Matplotlib package; and also map preparation using ETOPO1 relief, Shaded relief, Blue Marble etc. as background images with the help of basemap toolkit.

The focus of chapter 4 is on raster and vector data processing, which includes getting information of supported data formats by GDAL/OGR library, reading/writing of raster and vector data formats, commencing common tasks like preparing NDVI, contour generation, raster reprojection and resampling, and raster clipping.

Chapter 5 briefly describes libLAS library, which is used for reading and writing LIDAR data (ASPRS LAS format).

Chapter 6 deals with procuring some geospatial data which can be freely downloaded from the internet. Such data includes ASTER GDEM, SRTM data, GADM database, Natural Earth data etc.

The appendix provides all the scripts discussed in this literature.

1.12. TEST DATA

Following is a list of data which is used to demonstrate the functionality of scripts:

- *small_world.tiff* is a three band multi-spectral imagery of GeoTIFF format, which can be downloaded from *http://download.osgeo.org/*.
- *srtm_45_07.zip* file contains GeoTIFF format SRTM data downloaded from *http://srtm.csi.cgiar.org/*.
- *clipraster.zip* contains three band satellite imagery (*SatImage.tif*) of GeoTIFF format and a shapefile (*county.shp*) downloaded from *http://code.google.com/p/geospatialpython/downloads/list*.
- *p145r045_7t20011020_z43_nn30.tif.gz* and *p145r045_7t20011020_z43_nn40.tif.gz* are zipped LANDASAT imageries of a region (path: 145, row: 45) representing red band and NIR band. The imageries can be downloaded from *ftp://ftp.glcf.umd.edu/*.
- *LKA_adm0.shp* is a shapefile of administrative boundary of Sri Lanka, downloaded from *http://www.gadm.org/*.
- *Barrow_SeaIce_May7_2008.laz* is zipped LAS file, downloaded from *http://www.liblas.org/samples/*.

For demonstrative purpose, the test data (after unzipping) is placed in *C:\Data* folder, and the generated results will be kept in *C:\Results* folder.

BIBLIOGRAPHY

- ArcGIS Resources, *http://resources.arcgis.com/* [accessed on 05/07/2012].
- CGIAR-CSI SRTM database, *http://srtm.csi.cgiar.org/* [accessed on 10/04/2012].
- Global Land Cover Facility-University of Maryland (USA), *ftp://ftp.glcf.umd.edu/* [accessed on 18/08/2012].
- Google code, *http://code.google.com/* [accessed on 29/05/2012].
- IPython, *http://ipython.org/* [accessed on 20/07/2012].
- Open Geospatial Consortium, *http://www.opengeospatial.org/* [accessed on 15/05/2012].
- Open Source Geospatial Foundation, *http://www.osgeo.org/* [accessed on 03/09/2012].
- Open Source Initiative , *http://www.opensource.org/* [accessed on 01/01/2012].
- OSGeo Download Server, *http://download.osgeo.org/* [accessed on 30/08/2012].
- OSGeo Trac Instances, *http://trac.osgeo.org/* [accessed on 26/06/2012].
- Python Programming Language, *http://www.python.org/* [accessed on 05/07/2012].
- SciPy, *http://www.scipy.org/* [accessed on 10/05/2012].

Chapter 2
SPATIAL REFERENCING BY COORDINATES

Coordinates describing a position on or near the earth's surface are referenced to a model of the earth rather than to the earth itself. There are many models, and each model may be located with respect to the real earth in several different ways. The consequence is that one position on the real earth may be represented by multiple sets of coordinates, each referenced to different models. Furthermore, the direction, order and units of the coordinate system axes are subject to variation. Hence, without a set of geodetic parameters which identify the model and its relationship to the earth together with the coordinate system axes, coordinates are ambiguous.

Coordinate System is defined as a set of mathematical rules for specifying how coordinates are to be assigned to points. *Coordinate Reference System* (CRS), also referred as *Spatial Reference System* (SRS), is a coordinate system which is related to the real world by a datum.

2.1. EPSG DATASET

The *EPSG Geodetic Parameter Dataset*, abbreviated as *EPSG Dataset*, is a repository of parameters required to define a *Coordinate Reference System* (CRS), which ensures that coordinates describe the position unambiguously; and define coordinate operations that allow coordinates to be changed from one CRS to another. The *EPSG dataset* recognise the following types of coordinate operation:

- *Conversion*- A coordinate operation where both source and target CRS are based on the same datum. The most frequently encountered conversion is changing coordinates between geographic and projected.
- *Transformation*- A coordinate operation where source and target CRS are based on different datums.
- *Concatenated operation*- A series of transformations and/or conversions executed in sequence.

The CRSs held in the *EPSG Dataset* are typically those defined by national mapping organizations to be used for national mapping and spatial data infrastructures, as well as additional items of special interest to the petroleum exploration and production industry. The CRSs described are local, national, regional or global in extent. The dataset includes all components of these CRSs, for example datum, ellipsoid, prime meridian, map projection, coordinate system etc. The dataset also includes definitions for over 1500 coordinate transformations between CRSs. In *EPSG Dataset*, codes are assigned to CRSs, coordinate transformations, and their component entities (datums, projections, etc.).

The OGP Geomatics Committee, through its Geodesy Subcommittee, maintains and publishes the *EPSG Dataset*. The dataset is made available through three delivery mechanism:

- EPSG Registry- A web-based delivery platform (*http://www.epsg-registry.org/*).
- MS Access database.
- SQL scripts which enable a user to create an Oracle, MySQL, PostgreSQL or other database and populate that database with the EPSG Dataset.

2.2. PROJ.4 AND PYPROJ

One of the common challenge in geospatial data processing is conversion of coordinates from one system to another. PROJ.4 is a library which simplifies such task.

2.2.1. PROJ.4

PROJ.4 is a library capable of performing cartographic conversion/transformation and geodetic computation. PROJ.4 is an abbreviation for version 4 of the PROJ library. The library was originally written by Gerald Evenden, then of *United States Geological Survey* (USGS). PROJ.4 has been placed under the MIT license.

2.2.2. Pyproj

Pyproj is a wrapper which provides Python interfaces to PROJ.4 functions. Pyproj consists of only two classes, namely, *Proj* and *Geod*. Schematic representation of components of Pyproj library is shown in figure 2-1. The *Proj* class can convert from geographic coordinates (longitude, latitude) to projected coordinates (x,y) and vice versa; or from one projected coordinate system to another.

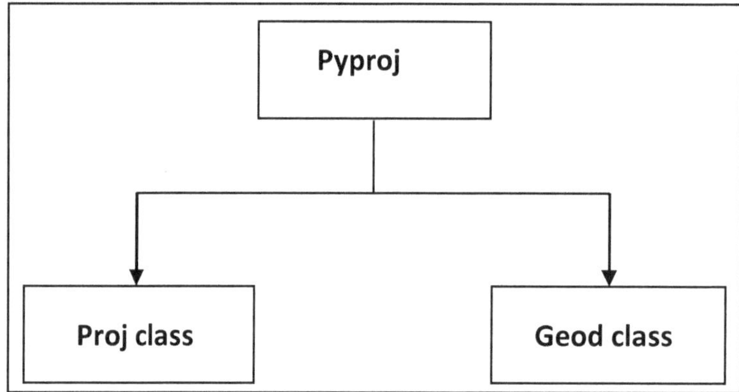

Figure 2-1: Components of Pyproj package

The *Geod* class can perform forward and inverse geodetic, or Great Circle computations. The forward computation involves determining latitude, longitude and back azimuth of a terminus point, given the latitude and longitude of an initial point, plus azimuth and distance. The inverse computation involves determining the forward and back azimuths and distance, given the latitudes and longitudes of an initial and terminus point (refer figure 2-2). In Pyproj library, input coordinates can be given as Python arrays, lists/tuples, scalars or NumPy arrays.

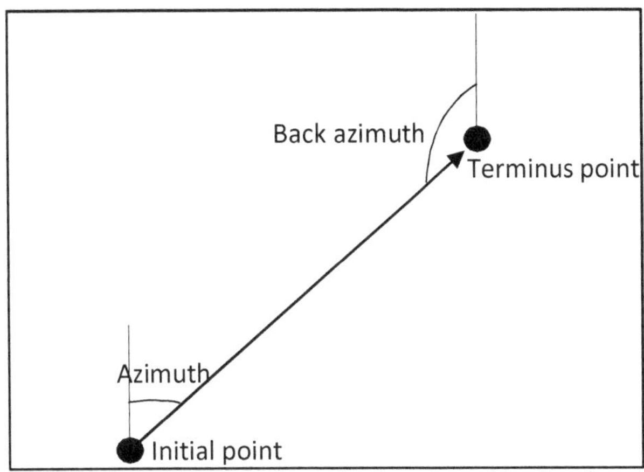

Figure 2-2: Representation of azimuth and back azimuth

2.2.3. Pyproj installation

By default, Pyproj is not included in Python(x,y) distribution, so one needs to download Microsoft Windows executable file of Pyproj (*pyproj-1.9.2.win32-py2.7.exe*) from *http://code.google.com/p/pyproj/* and install it in *Lib\site-packages* sub-folder of Python installation directory.

2.2.4. Pyproj information

Now write the first script with an aim to acquire some information about Pyproj library. There are many available Integrated Development Environments (IDE) to write a script; but as discussed previously, Spyder IDE is a good choice in this regard. The first line of script can be a shebang line. As shebang line is of no importance in Microsoft Windows environment, so one can ignore this line while writing scripts. But, inorder to generalize the architecture of script, shebang line will be included in all scripts of the book.

```
#!/usr/bin/env python
```

Import Pyproj library; print its version and its installation path. The module variables *__version__* and *__path__* are of string and list data types, respectively.

```
import pyproj
print 'Pyproj package version: ',pyproj.__version__
print 'Pyproj package installation path: ',pyproj.__path__
```

There is a data folder in Pyproj installation directory which contains support data for PROJ.4 programs as well as test scripts for testing cartographic projection program *proj* installation.

```
print 'PROJ.4 data path: ',pyproj.pyproj_datadir
```

The module variable *pj_list* is a dictionary containing all the supported projections and their descriptions. The information stored in *pj_list* has been tabulated in table 2-1.

```
print '\n\nSupported map projections (code::map projection name):'
for list in pyproj.pj_list:
   print '{0}::{1}'.format(list,pyproj.pj_list[list])
```

Table 2-1: Supported map projections by Pyproj

Map projection	Description
aea	Albers Equal Area
aeqd	Azimuthal Equidistant
airy	Airy
aitoff	Aitoff
alsk	Mod. Stererographics of Alaska
apian	Apian Globular I
august	August Epicycloidal
bacon	Bacon Globular
bipc	Bipolar conic of western hemisphere
boggs	Boggs Eumorphic
bonne	Bonne (Werner lat_1=90)
cass	Cassini
cc	Central Cylindrical
cea	Equal Area Cylindrical
chamb	Chamberlin Trimetric
collg	Collignon
crast	Craster Parabolic (Putnins P4)
denoy	Denoyer Semi-Elliptical
eck1	Eckert I
eck2	Eckert II
eck3	Eckert III
eck4	Eckert IV
eck5	Eckert V
eck6	Eckert VI
eqc	Equidistant Cylindrical (Plate Caree)
eqdc	Equidistant Conic
etmerc	Extended Transverse Mercator
euler	Euler
fahey	Fahey
fouc	Foucaut
fouc_s	Foucaut Sinusoidal
gall	Gall (Gall Stereographic)
geocent	Geocentric
geos	Geostationary Satellite View
gins8	Ginsburg VIII (TsNIIGAiK)
gn_sinu	General Sinusoidal Series
gnom	Gnomonic

Map projection	Description
goode	Goode Homolosine
gs48	Mod. Stererographics of 48 U.S.
gs50	Mod. Stererographics of 50 U.S.
hammer	Hammer & Eckert-Greifendorff
hatano	Hatano Asymmetrical Equal Area
healpix	HEALPix
rhealpix	rHEALPix
igh	Interrupted Goode Homolosine
imw_p	Internation Map of the World Polyconic
kav5	Kavraisky V
kav7	Kavraisky VII
krovak	Krovak
labrd	Laborde
laea	Lambert Azimuthal Equal Area
lagrng	Lagrange
larr	Larrivee
lask	Laskowski
lonlat	Lat/long (Geodetic)
latlon	Lat/long (Geodetic alias)
latlong	Lat/long (Geodetic alias)
longlat	Lat/long (Geodetic alias)
lcc	Lambert Conformal Conic
lcca	Lambert Conformal Conic Alternative
leac	Lambert Equal Area Conic
lee_os	Lee Oblated Stereographic
loxim	Loximuthal
lsat	Space oblique for LANDSAT
mbt_s	McBryde-Thomas Flat-Polar Sine
mbt_fps	McBryde-Thomas Flat-Pole Sine (No. 2)
mbtfpp	McBride-Thomas Flat-Polar Parabolic
mbtfpq	McBryde-Thomas Flat-Polar Quartic
mbtfps	McBryde-Thomas Flat-Polar Sinusoidal
merc	Mercator
mil_os	Miller Oblated Stereographic
mill	Miller Cylindrical
moll	Mollweide
murd1	Murdoch I
murd2	Murdoch II
murd3	Murdoch III
nell	Nell
nell_h	Nell-Hammer

Map projection	Description
nicol	Nicolosi Globular
nsper	Near-sided perspective
nzmg	New Zealand Map Grid
ob_tran	General Oblique Transformation
ocea	Oblique Cylindrical Equal Area
oea	Oblated Equal Area
omerc	Oblique Mercator
ortel	Ortelius Oval
ortho	Orthographic
pconic	Perspective Conic
poly	Polyconic (American)
putp1	Putnins P1
putp2	Putnins P2
putp3	Putnins P3
putp3p	Putnins P3
putp4p	Putnins P4
putp5	Putnins P5
putp5p	Putnins P5
putp6	Putnins P6
putp6p	Putnins P6
qua_aut	Quartic Authalic
robin	Robinson
rouss	Roussilhe Stereographic
rpoly	Rectangular Polyconic
sinu	Sinusoidal (Sanson-Flamsteed)
somerc	Swiss. Obl. Mercator
stere	Stereographic
sterea	Oblique Stereographic Alternative
gstmerc	Gauss-Schreiber Transverse Mercator (aka Gauss-Laborde Reunion)
tcc	Transverse Central Cylindrical
tcea	Transverse Cylindrical Equal Area
tissot	Tissot Conic
tmerc	Transverse Mercator
tpeqd	Two Point Equidistant
tpers	Tilted perspective
ups	Universal Polar Stereographic
urm5	Urmaev V
urmfps	Urmaev Flat-Polar Sinusoidal
utm	Universal Transverse Mercator (UTM)
vandg	van der Grinten (I)
vandg2	van der Grinten II

Map projection	Description
vandg3	van der Grinten III
vandg4	van der Grinten IV
vitk1	Vitkovsky I
wag1	Wagner I (Kavraisky VI)
wag2	Wagner II
wag3	Wagner III
wag4	Wagner IV
wag5	Wagner V
wag6	Wagner VI
wag7	Wagner VII
weren	Werenskiold I
wink1	Winkel I
wink2	Winkel II
wintri	Winkel Tripel

The module variable *pj_ellps* is a dictionary having supported ellipsoids (with their description) that are used in initializing *Geod* class object using *ellps* keyword argument. A comprehensive list of supported ellipsoids is given in table 2-2.

```
print '\n\nSupported ellipsoids (ellipsoid code::ellipsoid details):'
for list in pyproj.pj_ellps:
   print '{0}::{1}'.format(list,pyproj.pj_ellps[list])
```

Table 2-2: Supported ellipsoids by Pyproj

Ellipsoid	Parameter		Parameter		Description
airy	a	6377563.396	b	6356256.91	Airy 1830
andrae	a	6377104.43	rf	300	Andrae 1876 (Den., Iclnd.)
APL4.9	a	6378137	rf	298.25	Appl. Physics. 1965
aust_SA	a	6378160	rf	298.25	Australian Natl & S. Amer. 1969
bess_nam	a	6377483.865	rf	299.1528128	Bessel 1841 (Namibia)
bessel	a	6377397.155	rf	299.1528128	Bessel 1841
clrk66	a	6378206.4	b	6356583.8	Clarke 1866
clrk80	a	6378249.145	rf	293.4663	Clarke 1880 mod.
CPM	a	6375738.7	rf	334.29	Comm. des Poids et Mesures 1799
delmbr	a	6376428	rf	311.5	Delambre 1810 (Belgium)
engelis	a	6378136.05	rf	298.2566	Engelis 1985
evrst30	a	6377276.345	rf	300.8017	Everest 1830
evrst48	a	6377304.063	rf	300.8017	Everest 1948
evrst56	a	6377301.243	rf	300.8017	Everest 1956
evrst69	a	6377295.664	rf	300.8017	Everest 1969
evrstSS	a	6377298.556	rf	300.8017	Everest (Sabah & Sarawak)
fschr60	a	6378166	rf	298.3	Fischer (Mercury Datum) 1960

Ellipsoid	Parameter		Parameter		Description
fschr60m	a	6378155	rf	298.3	Modified Fischer 1960
fschr68	a	6378150	rf	298.3	Fischer 1968
GRS67	a	6378160	rf	298.2471674	GRS 67(IUGG 1967)
GRS80	a	6378137	rf	298.2572221	GRS 1980(IUGG, 1980)
helmert	a	6378200	rf	298.3	Helmert 1906
hough	a	6378270	rf	297	Hough
IAU76	a	6378140	rf	298.257	IAU 1976
intl	a	6378388	rf	297	International 1909 (Hayford)
kaula	a	6378163	rf	298.24	Kaula 1961
krass	a	6378245	rf	298.3	Krassovsky, 1942
lerch	a	6378139	rf	298.257	Lerch 1979
MERIT	a	6378137	rf	298.257	MERIT 1983
mod_airy	a	6377340.189	b	6356034.446	Modified Airy
mprts	a	6397300	rf	191	Maupertius 1738
new_intl	a	6378157.5	b	6356772.2	New International 1967
NWL9D	a	6378145	rf	298.25	Naval Weapons Lab., 1965
plessis	a	6376523	b	6355863	Plessis 1817 (France)
SEasia	a	6378155	b	6356773.321	Southeast Asia
SGS85	a	6378136	rf	298.257	Soviet Geodetic System 85
sphere	a	6370997	b	6370997	Normal Sphere
walbeck	a	6376896	b	6355834.847	Walbeck
WGS60	a	6378165	rf	298.3	WGS 60
WGS66	a	6378145	rf	298.25	WGS 66
WGS72	a	6378135	rf	298.26	WGS 72
WGS84	a	6378137	rf	298.2572236	WGS 84

2.2.5. Coordinate transformation

As discussed earlier, there is a *Proj* class in Pyproj package. The following script will demonstrate coordinate conversion approach and functionality of *Proj* class. The exercise will encompass conversion of location of Luanda (Angola) from geographic coordinate system to projected system, and vice-versa (results shown in figure 2-3). As discussed earlier, the first line is shebang line followed by importing Pyproj package.

```
#!/usr/bin/env python
import pyproj
```

Initialise a *Proj* class instance using PROJ.4 map projection control parameters key/value pairs. The key/value pairs can either be passed as a dictionary, or as PROJ.4 string, or as keyword arguments. Refer the PROJ.4 documentation (*http://trac.osgeo.org/proj/*) for more information about specifying projection parameters. Common usage of certain projections may be facilitated by the projection parameters which are pre-defined in initialisation files. These can be accessed by the parameter *+init=file:key*, where *file* is the filename (found at PROJ.4 data directory) containing the control information and *key* identifies the particular set of parameters in the file which needs to be included as projection parameters. If the optional keyword *preserve_units* is *True* (default is *False*), the units in map projection coordinates are not forced to be metre. The following commands define a *Proj* class instance (named *proj_wgs84*) having WGS84 spatial reference system, initialised using *epsg* and *esri* files, respectively.

```
proj_wgs84=pyproj.Proj(init='epsg:4326',preserve_units=True)
proj_wgs84=pyproj.Proj(init='esri:4326',preserve_units=True)
proj_wgs84=pyproj.Proj(projparams="+init=epsg:4326",preserve_units=True)
proj_wgs84=pyproj.Proj(projparams="+init=esri:4326",preserve_units=True)
```

Print the Proj.4 library version.

```
print 'PROJ version:',proj_wgs84.proj_version
```

Check, if the initialised *Proj* class instance has geographic coordinates using *is_latlong()* method; it returns *True* on success.

```
if proj_wgs84.is_latlong():
  print "EPSG:4326 coordinates are in lat/lon"
else:
  print "EPSG:4326 coordinates are not in lat/lon"
```

As Luanda lies in sount of zone 33 of UTM map projection of WGS84 spatial reference system, initialise a new *Proj* class instance (named *proj_wgs84_33s*) with WGS84 UTM 33S projection system as PROJ.4 string.

```
proj_wgs84_33s=pyproj.Proj(projparams='''+proj=utm +zone=33 +south +datum=WGS84
              +units=m''',preserve_units=True)
```

Assign geographic coordinates of a location in Luanda to variables.

```
lat=-8.852400;lon=13.237200
```

To convert coordinates from WGS84 geographic system to UTM projected system, one can use *transform()* method. It transforms a point between two coordinate systems defined by the *Proj* class instances *p1* and *p2*. The point *x1*, *y1*, *z1* in the coordinate system defined by *p1* is transformed to *x2*, *y2*, *z2* in the coordinate system defined by *p2*. *z1* is optional, if it is not set, it is assumed to be zero (and only *x2* and *y2* are returned). In addition of converting between cartographic and geographic projection coordinates, this function can take care of datum.

```
x2,y2=pyproj.transform(p1,p2,x1,y1)
```

If optional keyword *radians* is *True* (default is *False*), and if *p1* is defined in geographic coordinate (*p1.is_latlong()* is *True*); *x1, y1* is interpreted as radians instead of the default degrees. Similarly, if *p2* is defined in geographic coordinates and *radians=True*; *x2, y2* are returned in radians instead of degrees. If *p1.is_latlong()* and *p2.is_latlong()* both are *False*, the *radians* keyword has no effect. *x, y, z* can be NumPy or regular Python arrays, Python lists/tuples or scalars. For projections in geocentric coordinates, values of *x* and *y* are given in metre; *z* is always in metre.

```
x1,y1=pyproj.transform(p1=proj_wgs84,p2=proj_wgs84_33s,x=lon,y=lat,z=None, \
            radians=False)
```

Print the projected coordinates.

```
print 'Projected coordinates- X: %9.9f, Y: %9.9f'%(x1,y1)
```

This time, initialise the *Proj* class instance using keyword arguments.

```
proj_wgs84_33s=pyproj.Proj(proj='utm',zone=33,south=True,datum='WGS84', \
            preserve_units=True)
```

The Proj class instance can also be initialised using dictionary key/value pairs.

```
proj_wgs84_33s=pyproj.Proj(projparams={'proj':'utm','zone':33,'south':'True', \
            'datum':'WGS84','units':'m'},preserve_units=True)
```

Check if the *Proj* class instance has geocentric coordinate system or not. The *is_geocent()* method returns *True* if projection in geocentric (x, y) coordinates.

```
if proj_wgs84_33s.is_geocent():
  print "WGS 84 UTM 33S is geocentric coordinate system"
else:
  print "WGS 84 UTM 33S is not geocentric coordinate system"
```

Calling a *Proj* class instance with the arguments *lon,lat* will convert *lon/lat* (in degrees) to *x/y* native map projection coordinates (in metre). If optional keyword *inverse* is *True* (default is *False*), the inverse transformation from (*x, y*) to (*longitude, latitude*) is performed. If optional keyword *radians* is *True* (default is *False*), the units of *longitude, latitude* will be treated in radians instead of degrees. This works with NumPy and regular Python array objects, Python sequences and scalars.

```
lon,lat=proj_wgs84_33s(x1,y1,inverse=True,radians=False)
print 'Geographic coordinates- Longitude: %9.9f, Latitude: %9.9f' %(lon,lat)
```

Destroy Python objects by assigning them to *None*.

```
proj_wgs84=proj_wgs84_44n=None
```

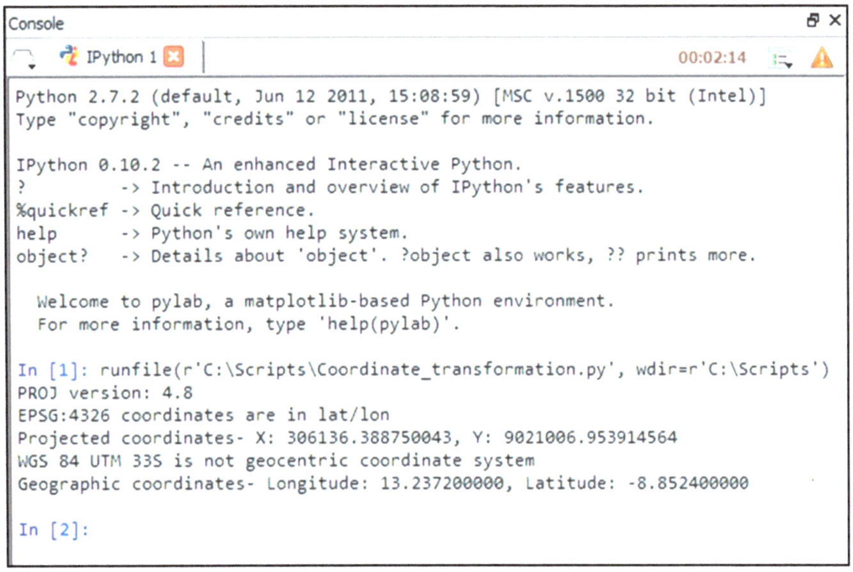

```
Console
IPython 1                                                00:02:14
Python 2.7.2 (default, Jun 12 2011, 15:08:59) [MSC v.1500 32 bit (Intel)]
Type "copyright", "credits" or "license" for more information.

IPython 0.10.2 -- An enhanced Interactive Python.
?         -> Introduction and overview of IPython's features.
%quickref -> Quick reference.
help      -> Python's own help system.
object?   -> Details about 'object'. ?object also works, ?? prints more.

  Welcome to pylab, a matplotlib-based Python environment.
  For more information, type 'help(pylab)'.

In [1]: runfile(r'C:\Scripts\Coordinate_transformation.py', wdir=r'C:\Scripts')
PROJ version: 4.8
EPSG:4326 coordinates are in lat/lon
Projected coordinates- X: 306136.388750043, Y: 9021006.953914564
WGS 84 UTM 33S is not geocentric coordinate system
Geographic coordinates- Longitude: 13.237200000, Latitude: -8.852400000

In [2]:
```

Figure 2-3: Coordinate transformation result

2.2.6. Geodetic computation

In this section, the discussion is on exploring *Geod* class functionalities of Pyproj package. The following script will deal with performing forward and inverse geodetic, and great circle computations (results shown in figure 2-4). Start with importing *Geod* class of Pyproj package.

#!/usr/bin/env python
from pyproj import Geod

Geod class instance can be initialised using either PROJ.4 geod initialisation string or by using keyword arguments. For both the approaches either one can directly specify the ellipsoids (refer table 2-2) that may be defined using the *ellps* parameter or by directly specifying the parameters of ellipsoid using *a* (semi-major or equatorial axis radius) keyword, and any one of the following keywords: *b* (semi-minor, or polar axis radius), *e* (eccentricity), *es* (eccentricity squared), *f* (flattening), or *rf* (reciprocal flattening).

g=Geod(initstring='+ellps=clrk66')
g=Geod(initstring='+a=6378206.4 +b=6356583.8')
The above syntaxes give an approach for initialising *Geod* class instance using PROJ.4 geod initialisation string while the following syntaxes is an keyword argument approach.

g=Geod(ellps='clrk66')
g=Geod(a=6378206.4,b=6356583.8)

Further, initialise variables with geographical coordinates of Portland and Boston cities.

boston_lat=42.+(15./60.); boston_lon=-71.-(7./60.)
portland_lat=45.+(31./60.); portland_lon=-123.-(41./60.)

Perform inverse transformation, which will returns forward and back azimuths, plus distances between initial points (specified by *lons1, lats1*) and terminus points (specified by *lons2, lats2*). The *inv* method works with NumPy and regular Python array objects, Python sequences and scalars. If optional keyword *radians=True* (*False* by default), longitude, latitude and azimuth will be considered as in radians instead of degrees. The distance is assumed to be in metre.

```
az12,az21,dist=g.inv(lons1=boston_lon,lats1=boston_lat,lons2=portland_lon, \
             lats2=portland_lat,radians=False)
print "Forward azimuth: %7.3f" %(az12)
print "Backward azimuth: %7.3f" %(az21)
print "Distance: %12.3f" % (dist)
```

Perform forward transformation, which will return longitudes, latitudes and back azimuths of terminus points corresponding to given longitudes (*lons1*) and latitudes (*lats1*) of initial points, plus forward azimuths (*az12*) and distances (*dist*). The *fwd* method works with NumPy and regular Python array objects, Python sequences and scalars. If *radians=True* (*False* by default), longitude, latitude and azimuths will be considered as in radians instead of degrees. The distance is assumed to be in metre.

```
endlon,endlat,backaz=g.fwd(lons=boston_lon,lats=boston_lat,az=az12,dist=dist, \
                radians=False)
print "Portland latitude: %7.3f" %(endlat)
print "Portland longitude: %7.3f" %(endlon)
print "Backward azimuth: %7.3f" % (backaz)
```

Get a list of longitude, latitude pairs which are equally spaced intermediate points along the geodesic between the given initial and terminus points (specified by Python floats *lon1, lat1* and *lon2, lat2*). If *radians=True*, longitude, latitude are considered in radians instead of degrees.

```
lonlats=g.npts(lon1=boston_lon,lat1=boston_lat,lon2=portland_lon, \
         lat2=portland_lat,npts=10,radians=False)
print 'Ten equally spaced points between Boston and Portland (lat, lon):'
for lon,lat in lonlats: print '%6.3f  %7.3f' %(lat,lon)
```

Destroy *g* by assigning it to *None*.

```
g=None
```

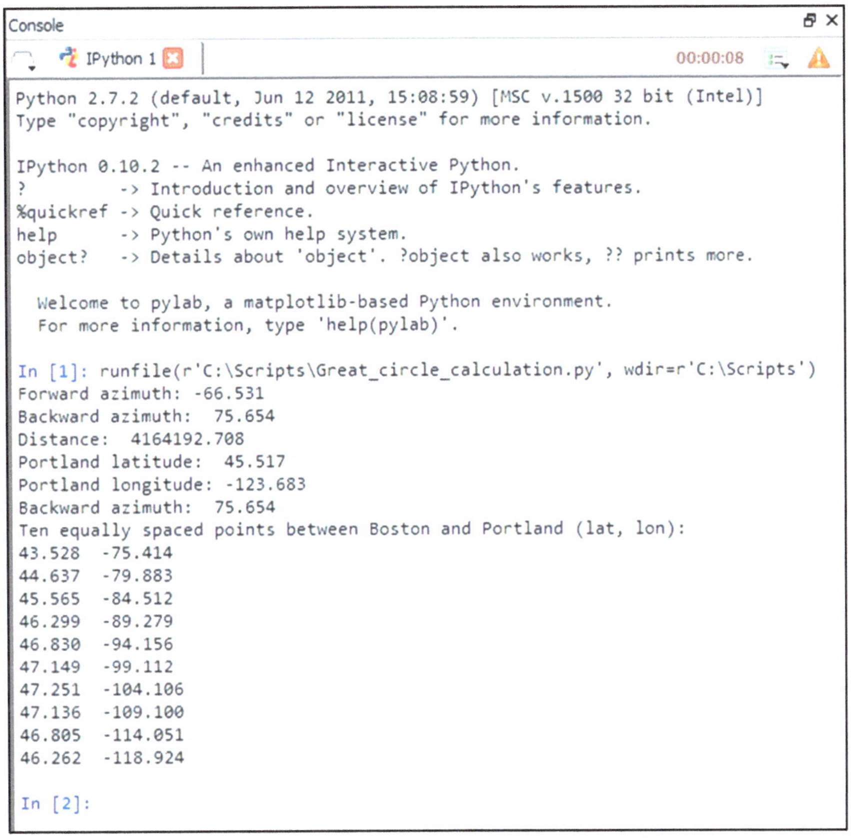

```
Console
IPython 1                                                    00:00:08

Python 2.7.2 (default, Jun 12 2011, 15:08:59) [MSC v.1500 32 bit (Intel)]
Type "copyright", "credits" or "license" for more information.

IPython 0.10.2 -- An enhanced Interactive Python.
?         -> Introduction and overview of IPython's features.
%quickref -> Quick reference.
help      -> Python's own help system.
object?   -> Details about 'object'. ?object also works, ?? prints more.

  Welcome to pylab, a matplotlib-based Python environment.
  For more information, type 'help(pylab)'.

In [1]: runfile(r'C:\Scripts\Great_circle_calculation.py', wdir=r'C:\Scripts')
Forward azimuth: -66.531
Backward azimuth:  75.654
Distance:  4164192.708
Portland latitude:  45.517
Portland longitude: -123.683
Backward azimuth:  75.654
Ten equally spaced points between Boston and Portland (lat, lon):
43.528  -75.414
44.637  -79.883
45.565  -84.512
46.299  -89.279
46.830  -94.156
47.149  -99.112
47.251  -104.106
47.136  -109.100
46.805  -114.051
46.262  -118.924

In [2]:
```

Figure 2-4: Great circle computation result

2.3. WKT REPRESENTATION OF SPATIAL REFERENCE SYSTEM

This topic discusses about Well-known Text (WKT) representation of SRS for geographic (latitude, longitude), projected (X,Y), and geocentric (X,Y,Z) coordinate system. For more details, refer the document *OpenGIS Implementation Standard for Geographic information- Simple feature access- Part 1: Common architecture* found at *http://www.opengeospatial.org/*. The WKT representation of SRS provides a standard textual representation for spatial reference system information. The coordinate system is composed of several objects, with each object has a keyword in upper case (for example, *DATUM* or *UNIT*) followed by comma-delimited parameters of the object in brackets. Some objects are composed of objects, so the result is a nested structure. Implementations are free to substitute standard brackets ( ) for square brackets [ ] and should be prepared to read both forms of brackets. The *Extended Backus Naur Form* (EBNF) definition for the string representation of a coordinate system is given in table 2-3, using square brackets.

Table 2-3: WKT representation of SRS

Keyword	Value
<spatial reference system> ::=	<projected cs> |
	<geographic cs> |
	<geocentric cs>
<projected cs> ::=	PROJCS <left delimiter>
	<csname>
	<comma> <geographic cs>

	<comma> <projection>
	(<comma> <parameter>)
	<comma> <linear unit>
	<right delimiter>
<geographic cs> ::=	GEOGCS <left delimiter> <csname>
	<comma> <datum>
	<comma> <prime meridian>
	<comma> <angular unit>
	(<comma> <linear unit>)
	<right delimiter>
<geocentric cs> ::=	GEOCCS <left delimiter>
	<name>
	<comma> <datum>
	<comma> <prime meridian>
	<comma> <linear unit>
	<right delimiter>
<datum> ::=	DATUM <left delimiter> <datum name>
	<comma> <spheroid>
	<right delimiter>
<projection> ::=	PROJECTION <left delimiter>
	<projection name>
	<right delimiter>
<parameter> ::=	PARAMETER <left delimiter>
	<parameter name>
	<comma> <value>
	<right delimiter>
<spheroid> ::=	SPHEROID <left delimiter>
	<spheroid name>
	<comma> <semi-major axis>
	<comma> <inverse flattening>
	<right delimiter>
<prime meridian> ::=	PRIMEM <left delimiter>
	<prime meridian name>
	<comma> <longitude>
	<right delimiter>
<linear unit> ::=	<unit>
<angular unit> ::=	<unit>
<unit> ::=	UNIT <left delimiter>
	<unit name>
	<comma> <conversion factor>
	<right delimiter>
<value> ::=	<signed numeric literal>

<semi-major axis> ::=	<signed numeric literal>
<longitude> ::=	<signed numeric literal>
<inverse flattening> ::=	<signed numeric literal>
<conversion factor> ::=	<signed numeric literal>
<unit name> ::=	<quoted name>
<spheroid name> ::=	<quoted name>
<projection name> ::=	<quoted name>
<prime meridian name> ::=	<quoted name>
<parameter name> ::=	<quoted name>
<datum name> ::=	<quoted name>
<csname> ::=	<quoted name>

Please note that the semi-major axis is measured in metre and shall be greater than zero. Conversion factor specifies number of metre (for a linear unit) or number of radians (for an angular unit) per unit and shall be greater than zero.

A data set's coordinate system is identified by the *PROJCS* keyword if the data are in projected coordinates, by *GEOGCS* if in geographic coordinates, or by *GEOCCS* if in geocentric coordinates. The *PROJCS* keyword is followed by all of the pieces which define the projected coordinate system. The first piece of any object is always the name. Several objects follow the projected coordinate system name: the geographic coordinate system, the map projection, zero or more parameters, and the linear unit of measure. All projected coordinate systems are based upon a geographic coordinate system, so the pieces specific to a projected coordinate system shall be described first. For example, the WKT representation of UTM Zone 10N on NAD83 datum is:

PROJCS["NAD_1983_UTM_Zone_10N",
GEOGCS["GCS_North_American_1983",
DATUM["D_North_American_1983",ELLIPSOID["GRS_1980",6378137,298.257222101]],
PRIMEM["Greenwich",0],UNIT["Degree",0.0174532925199433]],
PROJECTION["Transverse_Mercator"],PARAMETER["False_Easting",500000.0],
PARAMETER["False_Northing",0.0],PARAMETER["Central_Meridian",-123.0],
PARAMETER["Scale_Factor",0.9996],PARAMETER["Latitude_of_Origin",0.0], UNIT["Meter",1.0]]

2.4. SRS INFORMATION

Till now, SRS representation in the form of EPSG codes/dataset, PROJ.4, and WKT has been discussed. Now move further to explore how transformation of SRS representation from one format to another is carried out using GDAL/OGR library. For SRS related operations, one has to import OSR module from GDAL/OGR package (namespace *osgeo*). Apart from OSR module, there are four more modules that are included in the GDAL/OGR Python bindings, whose syntaxes for importing are as follows:

import osgeo.gdal as gdal
import osgeo.ogr as ogr
import osgeo.osr as osr

```
import osgeo.gdal_array as gdal_array
import osgeo.gdalconst as gdalconst
```

For importing a module from a package, *import package.item as itm* is preferred over the equivalent *from package import item*, because the latter is ambiguous as to whether *item* is a module, class or a method. The former makes it explicit that the programmer is importing a module or package. There are some modules whose names matches with the commonly used local variable names, e.g. *matplotlib.lines*, *matplotlib.colors* etc. To avoid clash, use *import package.item as itm* syntax.
Fetch list of supported projection methods using *GetProjectionMethods()* which are listed in table 2-4.

```
print "Supported map projections:"
for pm in osr.GetProjectionMethods():
    print pm[1]
```

Table 2-4: Supported map projections methods by GDAL/OGR

S. No.	Map projection methods
1	Transverse Mercator
2	Transverse Mercator (South Oriented)
3	Tunisia Mining Grid
4	Albers Conic Equal Area
5	Azimuthal Equidistant
6	Cylindrical Equal Area
7	Cassini/Soldner
8	Equidistant Conic
9	Bonne
10	Eckert I
11	Eckert II
12	Eckert III
13	Eckert IV
14	Eckert V
15	Eckert VI
16	Equirectangular
17	Gauss-Schreiber Transverse Mercator
18	Gall Stereographic
19	Goode Homolosine
20	Interrupted Goode Homolosine
21	Geostationary Satellite
22	Gnomonic
23	Hotine Oblique Mercator
24	Hotine Oblique Mercator Two Point Natural Origin
25	Lambert Azimuthal Equal Area
26	Lambert Conformal Conic (2SP)
27	Lambert Conformal Conic (1SP)

S. No.	Map projection methods
28	Lambert Conformal Conic (2SP - Belgium)
29	Miller Cylindrical
30	Mercator (1SP)
31	Mercator (2SP)
32	Mollweide
33	New Zealand Map Grid
34	Oblique Stereographic
35	Orthographic
36	Polyconic
37	Polar Stereographic
38	Robinson
39	Sinusoidal
40	Van Der Grinten
41	International Map of the World Polyconic
42	Wagner I (Kavraisky VI)
43	Wagner II
44	Wagner III
45	Wagner IV
46	Wagner V
47	Wagner VI
48	Wagner VII

Create two SRS instances, and initialise these based on EPSG GCS and PCS codes using *ImportFromEPSG()* method. The coordinate system definitions are normally read from the EPSG derived support files such as pcs.csv, gcs.csv, pcs.override.csv, gcs.override.csv and falling back to search for a PROJ.4 *epsg* init file. These support files are normally searched in the directory identified by the *GDAL_DATA* environemt variable. This method is relatively expensive, and generally involves quite a bit of text file scanning. Efforts should be made to avoid calling it many times for the same coordinate system.

```
SRS_GCS=osr.SpatialReference()
SRS_GCS.ImportFromEPSG(4326)
SRS_PCS=osr.SpatialReference()
SRS_PCS.ImportFromEPSG(32643)
```

Initialise a *CoordinateTransformation* class source and target SRS instances as arguments. Use *TransformPoint()* method to carry out coordinate transformation. If the argument to third parameter *z* is not provided, then the method takes zero by default.

```
xform=osr.CoordinateTransformation(SRS_GCS,SRS_PCS)
print "Coordinate transformation from GCS to PCS"
print xform.TransformPoint(75.867937,22.699574,0)
```

Check whether the defined SRS is geographic, projected, geocentric, compound, local or vertical coordinate system.

```
print "SRS_PCS description:"
if SRS_PCS.IsGeographic():
   print "SRS_PCS is a Geographic coordinate system."
if SRS_PCS.IsProjected():
   print "SRS_PCS is a Projected coordinate system."
if SRS_PCS.IsGeocentric():
   print "SRS_PCS is a Geocentric coordinate system."
if SRS_PCS.IsCompound():
   print "SRS_PCS is a Compound coordinate system."
if SRS_PCS.IsLocal():
   print "SRS_PCS is a Local coordinate system."
if SRS_PCS.IsVertical():
   print "SRS_PCS is a Vertical coordinate system."
```

Get UTM zone information using *GetUTMZone()*. The method returns a zone which is positive in northern hemisphere, negative in the southern hemisphere, and zero if SRS is not a UTM definition.

```
print "UTM Zone: ",SRS_PCS.GetUTMZone()
```

Fetch indicated attribute of named node using *GetAttrValue()* method. Thus, calling *GetAttrValue("UNIT",1)* would return the second child of the *UNIT* node, which is normally the length of the linear unit in metre. The method returns the requested value, or *None*, if it fails for any reason.

```
print "Projected coordinate system: ",SRS_PCS.GetAttrValue("PROJCS",0)
print "Geographic coordinate system: ",SRS_PCS.GetAttrValue("GEOGCS",0)
print "Projection: ",SRS_PCS.GetAttrValue("PROJECTION",0)
print "Datum: ",SRS_PCS.GetAttrValue("DATUM",0)
print "Spheroid: ",SRS_PCS.GetAttrValue("SPHEROID",0)
```

Get spheroid's semi major axis, semi minor axis and inverse flattening.

```
print "Semi major axis: ",SRS_PCS.GetSemiMajor()
print "Semi minor axis: ",SRS_PCS.GetSemiMinor()
print "Inverse flattening: ",SRS_PCS.GetInvFlattening()
```

Get first attribute of PRIMEM (Prime meridian) node.

```
print "Prime Meridian: ",SRS_PCS.GetAttrValue("PRIMEM",0)
```

Fetch some projection parameters value using *GetProjParm()* method.

```
print "Projection parameter (Latitude of origin): ",SRS_PCS.GetProjParm("latitude_of_origin")
print "Projection parameter (Central meridian): ",SRS_PCS.GetProjParm("central_meridian")
print "Projection parameter (Scale factor): ",SRS_PCS.GetProjParm("scale_factor")
print "Projection parameter (False easting): ",SRS_PCS.GetProjParm("false_easting")
print "Projection parameter (False northing): ",SRS_PCS.GetProjParm("false_northing")
```

Fetch linear projection unit, which is a value to multiply by linear distances to transform them to metre. *GetLinearUnitsName()* and *GetLinearUnits()* methods checks directly under the PROJCS, GEOCCS or LOCAL_CS node for units.

```
print "Linear unit name: ",SRS_PCS.GetLinearUnitsName()
print "Linear unit: ",SRS_PCS.GetLinearUnits()
```

Get angular unit name.

```
print "Angular unit name: ",SRS_PCS.GetAttrValue("GEOGCS|UNIT",0)
```

Fetch angular unit of geographic coordinate system, which is the value to multiply by angular distances to transform them to radians. *GetAngularUnits()* method checks directly under the *GEOGCS* node for units.

```
print "Angular unit: ",SRS_PCS.GetAngularUnits()
```

Query *AUTHORITY[]* node from within the WKT tree, and fetch the authority name and code value. The most common authority is *EPSG*. The *GetAuthorityName()* and *GetAuthorityCode()* methods takes argument which is a partial or complete path to the node to get an authority from. ie. "PROJCS", "GEOGCS", "GEOGCS|UNIT" etc. These methods return the name code and value code from authority node, or *None* on failure.

```
print "Authority name: ",SRS_PCS.GetAuthorityName("PROJCS")
print "Authority code: ",SRS_PCS.GetAuthorityCode("PROJCS")
```

Convert the SRS into WKT format.

```
print "WKT format: \n",SRS_PCS.ExportToWkt()
```

For display purpose, convert the SRS into a a nicely formatted WKT string. If the argument in *ExportToPrettyWkt()* is 0 (default), then the *AXIS*, *AUTHORITY* and *EXTENSION* nodes will not be stripped off.

```
print "Pretty WKT format: \n",SRS_PCS.ExportToPrettyWkt(0)
```

Converts the SRS into PROJ.4 format.

```
print "Proj.4 format: \n",SRS_PCS.ExportToProj4()
```

Convert the SRS to ESRI WKT format. The value nodes of this coordinate system are modified in various manners to more closely map onto the ESRI concept of WKT format. This includes renaming a variety of projections and arguments, and stripping out nodes note recognised by ESRI (like *AUTHORITY* and *AXIS*).

```
SRS_PCS.MorphToESRI()
print "ESRI compatible WKT for use as *.prj: \n", SRS_PCS.ExportToWkt()
```

Export SRS into MapInfo style coordinate system format.

```
print "MapInfo format: \n",SRS_PCS.ExportToMICoordSys()
```

Destroy the OSR instances by assigning *None*.

```
SRS_GCS=SRS_PCS=None
```

This chapter provided a glimpse of spatial reference system, coordinate transformation and great circle computation, which is an important domain of geoinformatics.

BIBLIOGRAPHY

- __init__.py, Pyproj 1.9.0 package.
- osr.py, GDAL/OGR 1.9.1 package.
- Environmental Systems Research Institute, *http://www.esri.com/* [accessed on 10/08/2012].
- European Petroleum Survey Group, *http://www.epsg.org/* [accessed on 02/07/2012].
- FTP alias for OSGeo Download Server, *ftp://ftp.remotesensing.org/* [accessed on 10/06/2012].
- Geospatial Data Abstraction Library, *http://www.gdal.org/* [accessed on 09/09/2012].
- Google code, *http://code.google.com/* [accessed on 05/06/2012].
- Open Geospatial Consortium, *http://www.opengeospatial.org/* [accessed on 016/06/2012].
- OSGeo Trac Instances, *http://trac.osgeo.org/* [accessed on 08/08/2012].

Chapter 3
CARTOGRAPHY

In order to represent the curved surface of the earth on a two-dimensional map, a map projection is needed. Since this cannot be done without distortion, there are many map projections, each with its own advantages and disadvantages.

3.1. MATPLOTLIB

Matplotlib is a plotting library for Python programming language which produces publication quality figures. Although Matplotlib is written primarily in Python, it makes heavy use of NumPy and other extension code to provide good performance, even for large arrays. Matplotlib is a whole package; and it contains a module named Pyplot. Pyplot provides a MATLAB style interface to the underlying plotting library in Matplotlib. The functionality of Matplotlib extends with the inclusion of application-specific functions in the form of toolkits, namely, Basemap, GTK tools, Excel tools, Natgrid, mplot3d, and AxesGrid.

3.2. BASEMAP TOOLKIT

The Matplotlib Basemap toolkit is a library for 2D map data plotting in Python. Basemap does not do any plotting on its own, but provides the facilities to transform coordinates to one of the supported map projections using the PROJ.4 library. Matplotlib is then used to plot contours, images, vectors, shoreline, river, political boundary etc. in the transformed coordinates. Basemap also provides facilities for reading shapefiles. This chapter focuses on preliminary information on map plotting using Matplotlib, so the reader of this book should refer to some other literature source for detailed information on Matplotlib methods and classes.

3.2.1. Installation

By default, Matplotlib is included in Python(x,y) distribution, but some toolkits are not packaged by default; so one needs to independently download and install them. The toolkits have a different namespace *mpl_toolkits*, and they do not get installed in the default Matplotlib installation directory. Pre-built executable file of Basemap toolkit (*basemap-1.0.5.win32-py2.7.exe*) can be obtained from the link: *http://sourceforge.net/projects/matplotlib/files/matplotlib-toolkits/*.

3.3. BASEMAP INFORMATION

Try to obtain some basic information, such as, Matplotlib and Basemap toolkit version, supported map projections and projection parameters.

```
#!/usr/bin/env python
import matplotlib
print "Matplotlib version: ",matplotlib.__version__
```

One can also know the version of Matplotlib by importing Basemap toolkit (for plotting map data with the assistance Matplotlib).

#!/usr/bin/env python
import mpl_toolkits.basemap as basemap
print "Matplotlib version: ",basemap._matplotlib_version

Print installed versions of Basemap toolkit.

print "Matplotlib basemap toolkit version: ",basemap.__version__

Get directory path of Basemap toolkit data files (*lib\site-packages\mpl_toolkits\basemap\data*).

print "Matplotlib basemap toolkit data directory: ",basemap.basemap_datadir

Basemap toolkit contains *Basemap* class, which does most of the map making work. The module variable *supported_projections* is a string which contains short name of map projection (used with the projection keyword to define a projection when creating a *Basemap* class instance) and its corresponding descriptive name, given in table 3-1.

Table 3-1: Supported map projection

Code	Map projection
aea	Albers Equal Area
aeqd	Azimuthal Equidistant
cass	Cassini-Soldner
cyl	Cylindrical Equidistant
eck4	Eckert IV
eqdc	Equidistant Conic
gall	Gall Stereographic Cylindrical
geos	Geostationary
gnom	Gnomonic
hammer	Hammer
kav7	Kavrayskiy VII
laea	Lambert Azimuthal Equal Area
lcc	Lambert Conformal
mbtfpq	McBryde-Thomas Flat-Polar Quartic
merc	Mercator
mill	Miller Cylindrical
moll	Mollweide
npaeqd	North-Polar Azimuthal Equidistant
nplaea	North-Polar Lambert Azimuthal
npstere	North-Polar Stereographic
nsper	Near-Sided Perspective
omerc	Oblique Mercator

Code	Map projection
ortho	Orthographic
poly	Polyconic
robin	Robinson
sinu	Sinusoidal
spaeqd	South-Polar Azimuthal Equidistant
splaea	South-Polar Lambert Azimuthal
spstere	South-Polar Stereographic
stere	Stereographic
tmerc	Transverse Mercator
vandg	Van der Grinten

```
print "\nSupported map projections:\n",basemap.supported_projections
```

Note that many map projection possess one of two desirable properties- either they can be of equal-area (the area of features is preserved) or conformal (the shape of features is preserved). Since no map projection can have both at the same time, many compromise between the two.

The module variable *projection_params* is a dictionary that provides parameters which can be used to define the properties of each map projection (refer table 3-2).

```
print 'Supported map projection parameters (code:projection parameters):'
for prj_param in basemap.projection_params:
   print '{0}\t: {1}'.format(prj_param,basemap.projection_params[prj_param])
```

Table 3-2: Map projection parameters

Code	Map projection parameters
aea	lon_0,lat_0,lat_1
aeqd	lon_0,lat_0
cass	lon_0,lat_0
cyl	corners only (no width/height)
eck4	lon_0,lat_0,no corners or width/height
eqdc	lon_0,lat_0,lat_1,lat_2
gall	corners only (no width/height)
geos	lon_0,satellite_height,llcrnrx,llcrnry,urcrnrx,urcrnry,no width/height
gnom	lon_0,lat_0
hammer	lon_0,lat_0,no corners or width/height
kav7	lon_0,lat_0,no corners or width/height
laea	lon_0,lat_0
lcc	lon_0,lat_0,lat_1,lat_2
mbtfpq	lon_0,lat_0,no corners or width/height
merc	corners plus lat_ts (no width/height)
mill	corners only (no width/height)

moll	lon_0,lat_0,no corners or width/height
npaeqd	bounding_lat,lon_0,lat_0,no corners or width/height
nplaea	bounding_lat,lon_0,lat_0,no corners or width/height
npstere	bounding_lat,lon_0,lat_0,no corners or width/height
nsper	lon_0,satellite_height,llcrnrx,llcrnry,urcrnrx,urcrnry,no width/height
omerc	lon_0,lat_0,lat_1,lat_2,lon_1,lon_2,no_rot
ortho	lon_0,lat_0,llcrnrx,llcrnry,urcrnrx,urcrnry,no width/height
poly	lon_0,lat_0
robin	lon_0,lat_0,no corners or width/height
sinu	lon_0,lat_0,no corners or width/height
spaeqd	bounding_lat,lon_0,lat_0,no corners or width/height
splaea	bounding_lat,lon_0,lat_0,no corners or width/height
spstere	bounding_lat,lon_0,lat_0,no corners or width/height
stere	lon_0,lat_0,lat_ts
tmerc	lon_0,lat_0
vandg	lon_0,lat_0,no corners or width/height

Map projection parameters (given in table 3.2) will be discussed in later section of this chapter. Print cylindrical (cyl, merc, mill, and gall) and pseudocylindrical (moll, robin, eck4, kav7, sinu, mbtfpq, vandg, and hammer) map projections.

```
print "\nCylindrical projections: ",basemap._cylproj
print "Pseudocylindrical projections: ",basemap._pseudocyl
```

3.4. SUPPORTED COLOURS AND COLOURMAPS

Matplotlib supports various approaches to specify colours information. For the basic built-in colours, one can use a single letter (given in table 3-3). The following patch of code will print the built-in colour information.

```
#!/usr/bin/env python
import matplotlib.colors as colors
print "Basic built-in colours (code:RGB tuple)"
for clr in colors.ColorConverter.colors:
   print '{0}: {1}'.format(clr,colors.ColorConverter.colors[clr])
```

Table 3-3: Colour alias

Alias	Colour	RGB tuple
b	Blue	(0.0, 0.0, 1.0)
c	Cyan	(0.0, 0.75, 0.75)
g	Green	(0.0, 0.5, 0.0)
k	Black	(0.0, 0.0, 0.0)
m	Magenta	(0.75, 0, 0.75)
r	Red	(1.0, 0.0, 0.0)

Alias	Colour	RGB tuple
w	White	(1.0, 1.0, 1.0)
y	Yellow	(0.75, 0.75, 0)

Gray shades can be given as a string, encoding a float in the range [0,1], e.g.

colour='0.75'

For a greater range of colours, there are two options; either specify the color using an HTML hex string, as in:
colour='#eeefff'

or you can pass an RGB tuple, where each of R , G , B are in the range [0,1]:

colour=[0.1,0.2,0.8]

Finally, legal HTML names for colours, like "red", "burlywood" and "chartreuse" are supported; a comprehensive list is given in table 3-4. The following patch of code will return colour name, its hex code and its corresponding RGB tuple.

```
print "\nAll supported colours (colour:Hex string:RGB tuple)"
for clr in colors.cnames:
   HEX=colors.cnames[clr]
   print '{0}:{1}:{2}'.format(clr,HEX,colors.hex2color(HEX))
```

Table 3-4: Colour names with hexadecimal code and RGB tuple

Colour name	Hex	RGB tuple
aliceblue	#F0F8FF	(0.9411764705882353, 0.9725490196078431, 1.0)
antiquewhite	#FAEBD7	(0.9803921568627451, 0.9215686274509803, 0.8431372549019608)
aqua	#00FFFF	(0.0, 1.0, 1.0)
aquamarine	#7FFFD4	(0.4980392156862745, 1.0, 0.8313725490196079)
azure	#F0FFFF	(0.9411764705882353, 1.0, 1.0)
beige	#F5F5DC	(0.9607843137254902, 0.9607843137254902, 0.8627450980392157)
bisque	#FFE4C4	(1.0, 0.8941176470588236, 0.7686274509803922)
black	#000000	(0.0, 0.0, 0.0)
blanchedalmond	#FFEBCD	(1.0, 0.9215686274509803, 0.803921568627451)
blue	#0000FF	(0.0, 0.0, 1.0)
blueviolet	#8A2BE2	(0.5411764705882353, 0.16862745098039217, 0.8862745098039215)
brown	#A52A2A	(0.6470588235294118, 0.16470588235294117, 0.16470588235294117)
burlywood	#DEB887	(0.8705882352941177, 0.7215686274509804, 0.5294117647058824)
cadetblue	#5F9EA0	(0.37254901960784315, 0.6196078431372549, 0.6274509803921569)
chartreuse	#7FFF00	(0.4980392156862745, 1.0, 0.0)
chocolate	#D2691E	(0.8235294117647058, 0.4117647058823529, 0.11764705882352941)
coral	#FF7F50	(1.0, 0.4980392156862745, 0.3137254901960784)
cornflowerblue	#6495ED	(0.39215686274509803, 0.5843137254901961, 0.9294117647058824)

Colour name	Hex	RGB tuple
cornsilk	#FFF8DC	(1.0, 0.9725490196078431, 0.8627450980392157)
crimson	#DC143C	(0.8627450980392157, 0.0784313725490196, 0.23529411764705882)
cyan	#00FFFF	(0.0, 1.0, 1.0)
darkblue	#00008B	(0.0, 0.0, 0.5450980392156862)
darkcyan	#008B8B	(0.0, 0.5450980392156862, 0.5450980392156862)
darkgoldenrod	#B8860B	(0.7215686274509804, 0.5254901960784314, 0.043137254901960784)
darkgray	#A9A9A9	(0.6627450980392157, 0.6627450980392157, 0.6627450980392157)
darkgreen	#006400	(0.0, 0.39215686274509803, 0.0)
darkgrey	#A9A9A9	(0.6627450980392157, 0.6627450980392157, 0.6627450980392157)
darkkhaki	#BDB76B	(0.7411764705882353, 0.7176470588235294, 0.4196078431372549)
darkmagenta	#8B008B	(0.5450980392156862, 0.0, 0.5450980392156862)
darkolivegreen	#556B2F	(0.3333333333333333, 0.4196078431372549, 0.1843137254901961)
darkorange	#FF8C00	(1.0, 0.5490196078431373, 0.0)
darkorchid	#9932CC	(0.6, 0.19607843137254902, 0.8)
darkred	#8B0000	(0.5450980392156862, 0.0, 0.0)
darksalmon	#E9967A	(0.9137254901960784, 0.5882352941176471, 0.47843137254901963)
darkseagreen	#8FBC8F	(0.5607843137254902, 0.7372549019607844, 0.5607843137254902)
darkslateblue	#483D8B	(0.2823529411764706, 0.23921568627450981, 0.5450980392156862)
darkslategray	#2F4F4F	(0.1843137254901961, 0.30980392156862746, 0.30980392156862746)
darkslategrey	#2F4F4F	(0.1843137254901961, 0.30980392156862746, 0.30980392156862746)
darkturquoise	#00CED1	(0.0, 0.807843137254902, 0.8196078431372549)
darkviolet	#9400D3	(0.5803921568627451, 0.0, 0.8274509803921568)
deeppink	#FF1493	(1.0, 0.0784313725490196, 0.5764705882352941)
deepskyblue	#00BFFF	(0.0, 0.7490196078431373, 1.0)
dimgray	#696969	(0.4117647058823529, 0.4117647058823529, 0.4117647058823529)
dimgrey	#696969	(0.4117647058823529, 0.4117647058823529, 0.4117647058823529)
dodgerblue	#1E90FF	(0.11764705882352941, 0.5647058823529412, 1.0)
firebrick	#B22222	(0.6980392156862745, 0.13333333333333333, 0.13333333333333333)
floralwhite	#FFFAF0	(1.0, 0.9803921568627451, 0.9411764705882353)
forestgreen	#228B22	(0.13333333333333333, 0.5450980392156862, 0.13333333333333333)
fuchsia	#FF00FF	(1.0, 0.0, 1.0)
gainsboro	#DCDCDC	(0.8627450980392157, 0.8627450980392157, 0.8627450980392157)
ghostwhite	#F8F8FF	(0.9725490196078431, 0.9725490196078431, 1.0)
gold	#FFD700	(1.0, 0.8431372549019608, 0.0)
goldenrod	#DAA520	(0.8549019607843137, 0.6470588235294118, 0.12549019607843137)
gray	#808080	(0.5019607843137255, 0.5019607843137255, 0.5019607843137255)
green	#008000	(0.0, 0.5019607843137255, 0.0)
greenyellow	#ADFF2F	(0.6784313725490196, 1.0, 0.1843137254901961)
grey	#808080	(0.5019607843137255, 0.5019607843137255, 0.5019607843137255)
honeydew	#F0FFF0	(0.9411764705882353, 1.0, 0.9411764705882353)
hotpink	#FF69B4	(1.0, 0.4117647058823529, 0.7058823529411765)
indianred	#CD5C5C	(0.803921568627451, 0.3607843137254902, 0.3607843137254902)
indigo	#4B0082	(0.29411764705882354, 0.0, 0.5098039215686274)
ivory	#FFFFF0	(1.0, 1.0, 0.9411764705882353)
khaki	#F0E68C	(0.9411764705882353, 0.9019607843137255, 0.5490196078431373)

Colour name	Hex	RGB tuple
lavender	#E6E6FA	(0.9019607843137255, 0.9019607843137255, 0.9803921568627451)
lavenderblush	#FFF0F5	(1.0, 0.9411764705882353, 0.9607843137254902)
lawngreen	#7CFC00	(0.48627450980392156, 0.9882352941176471, 0.0)
lemonchiffon	#FFFACD	(1.0, 0.9803921568627451, 0.803921568627451)
lightblue	#ADD8E6	(0.6784313725490196, 0.8470588235294118, 0.9019607843137255)
lightcoral	#F08080	(0.9411764705882353, 0.5019607843137255, 0.5019607843137255)
lightcyan	#E0FFFF	(0.8784313725490196, 1.0, 1.0)
lightgoldenrodyellow	#FAFAD2	(0.9803921568627451, 0.9803921568627451, 0.8235294117647058)
lightgray	#D3D3D3	(0.8274509803921568, 0.8274509803921568, 0.8274509803921568)
lightgreen	#90EE90	(0.5647058823529412, 0.9333333333333333, 0.5647058823529412)
lightgrey	#D3D3D3	(0.8274509803921568, 0.8274509803921568, 0.8274509803921568)
lightpink	#FFB6C1	(1.0, 0.7137254901960784, 0.7568627450980392)
lightsalmon	#FFA07A	(1.0, 0.6274509803921569, 0.47843137254901963)
lightseagreen	#20B2AA	(0.12549019607843137, 0.6980392156862745, 0.6666666666666666)
lightskyblue	#87CEFA	(0.5294117647058824, 0.807843137254902, 0.9803921568627451)
lightslategray	#778899	(0.4666666666666667, 0.5333333333333333, 0.6)
lightslategrey	#778899	(0.4666666666666667, 0.5333333333333333, 0.6)
lightsteelblue	#B0C4DE	(0.6901960784313725, 0.7686274509803922, 0.8705882352941177)
lightyellow	#FFFFE0	(1.0, 1.0, 0.8784313725490196)
lime	#00FF00	(0.0, 1.0, 0.0)
limegreen	#32CD32	(0.19607843137254902, 0.803921568627451, 0.19607843137254902)
linen	#FAF0E6	(0.9803921568627451, 0.9411764705882353, 0.9019607843137255)
magenta	#FF00FF	(1.0, 0.0, 1.0)
maroon	#800000	(0.5019607843137255, 0.0, 0.0)
mediumaquamarine	#66CDAA	(0.4, 0.803921568627451, 0.6666666666666666)
mediumblue	#0000CD	(0.0, 0.0, 0.803921568627451)
mediumorchid	#BA55D3	(0.7294117647058823, 0.3333333333333333, 0.8274509803921568)
mediumpurple	#9370DB	(0.5764705882352941, 0.4392156862745098, 0.8588235294117647)
mediumseagreen	#3CB371	(0.23529411764705882, 0.7019607843137254, 0.44313725490196076)
mediumslateblue	#7B68EE	(0.4823529411764706, 0.40784313725490196, 0.9333333333333333)
mediumspringgreen	#00FA9A	(0.0, 0.9803921568627451, 0.6039215686274509)
mediumturquoise	#48D1CC	(0.2823529411764706, 0.8196078431372549, 0.8)
mediumvioletred	#C71585	(0.7803921568627451, 0.08235294117647059, 0.5215686274509804)
midnightblue	#191970	(0.09803921568627451, 0.09803921568627451, 0.4392156862745098)
mintcream	#F5FFFA	(0.9607843137254902, 1.0, 0.9803921568627451)
mistyrose	#FFE4E1	(1.0, 0.8941176470588236, 0.8823529411764706)
moccasin	#FFE4B5	(1.0, 0.8941176470588236, 0.7098039215686275)
navajowhite	#FFDEAD	(1.0, 0.8705882352941177, 0.6784313725490196)
navy	#000080	(0.0, 0.0, 0.5019607843137255)
oldlace	#FDF5E6	(0.9921568627450981, 0.9607843137254902, 0.9019607843137255)
olive	#808000	(0.5019607843137255, 0.5019607843137255, 0.0)
olivedrab	#6B8E23	(0.4196078431372549, 0.5568627450980392, 0.13725490196078433)
orange	#FFA500	(1.0, 0.6470588235294118, 0.0)
orangered	#FF4500	(1.0, 0.27058823529411763, 0.0)
orchid	#DA70D6	(0.8549019607843137, 0.4392156862745098, 0.8392156862745098)

Colour name	Hex	RGB tuple
palegoldenrod	#EEE8AA	(0.9333333333333333, 0.9098039215686274, 0.6666666666666666)
palegreen	#98FB98	(0.596078431372549, 0.984313725490196, 0.596078431372549)
palevioletred	#AFEEEE	(0.6862745098039216, 0.9333333333333333, 0.9333333333333333)
papayawhip	#FFEFD5	(1.0, 0.9372549019607843, 0.8352941176470589)
peachpuff	#FFDAB9	(1.0, 0.8549019607843137, 0.7254901960784313)
peru	#CD853F	(0.803921568627451, 0.5215686274509804, 0.24705882352941178)
pink	#FFC0CB	(1.0, 0.7529411764705882, 0.796078431372549)
plum	#DDA0DD	(0.8666666666666667, 0.6274509803921569, 0.8666666666666667)
powderblue	#B0E0E6	(0.6901960784313725, 0.8784313725490196, 0.9019607843137255)
purple	#800080	(0.5019607843137255, 0.0, 0.5019607843137255)
red	#FF0000	(1.0, 0.0, 0.0)
rosybrown	#BC8F8F	(0.7372549019607844, 0.5607843137254902, 0.5607843137254902)
royalblue	#4169E1	(0.2549019607843137, 0.4117647058823529, 0.8823529411764706)
saddlebrown	#8B4513	(0.5450980392156862, 0.27058823529411763, 0.07450980392156863)
salmon	#FA8072	(0.9803921568627451, 0.5019607843137255, 0.4470588235294118)
sandybrown	#FAA460	(0.9803921568627451, 0.6431372549019608, 0.3764705882352941)
seagreen	#2E8B57	(0.1803921568627451, 0.5450980392156862, 0.3411764705882353)
seashell	#FFF5EE	(1.0, 0.9607843137254902, 0.9333333333333333)
sienna	#A0522D	(0.6274509803921569, 0.3215686274509804, 0.17647058823529413)
silver	#C0C0C0	(0.7529411764705882, 0.7529411764705882, 0.7529411764705882)
skyblue	#87CEEB	(0.5294117647058824, 0.807843137254902, 0.9215686274509803)
slateblue	#6A5ACD	(0.41568627450980394, 0.35294117647058826, 0.803921568627451)
slategray	#708090	(0.4392156862745098, 0.5019607843137255, 0.5647058823529412)
slategrey	#708090	(0.4392156862745098, 0.5019607843137255, 0.5647058823529412)
snow	#FFFAFA	(1.0, 0.9803921568627451, 0.9803921568627451)
springgreen	#00FF7F	(0.0, 1.0, 0.4980392156862745)
steelblue	#4682B4	(0.27450980392156865, 0.5098039215686274, 0.7058823529411765)
tan	#D2B48C	(0.8235294117647058, 0.7058823529411765, 0.5490196078431373)
teal	#008080	(0.0, 0.5019607843137255, 0.5019607843137255)
thistle	#D8BFD8	(0.8470588235294118, 0.7490196078431373, 0.8470588235294118)
tomato	#FF6347	(1.0, 0.38823529411764707, 0.2784313725490196)
turquoise	#40E0D0	(0.25098039215686274, 0.8784313725490196, 0.8156862745098039)
violet	#EE82EE	(0.9333333333333333, 0.5098039215686274, 0.9333333333333333)
wheat	#F5DEB3	(0.9607843137254902, 0.8705882352941177, 0.7019607843137254)
white	#FFFFFF	(1.0, 1.0, 1.0)
whitesmoke	#F5F5F5	(0.9607843137254902, 0.9607843137254902, 0.9607843137254902)
yellow	#FFFF00	(1.0, 1.0, 0.0)
yellowgreen	#9ACD32	(0.6039215686274509, 0.803921568627451, 0.19607843137254902)

Figure 3-1 shows the available colourmaps in Matplotlib. Some of the colourmaps are ignored, which are basically the ones that are simply reversed versions of ones that are depicted in figure 3-1.

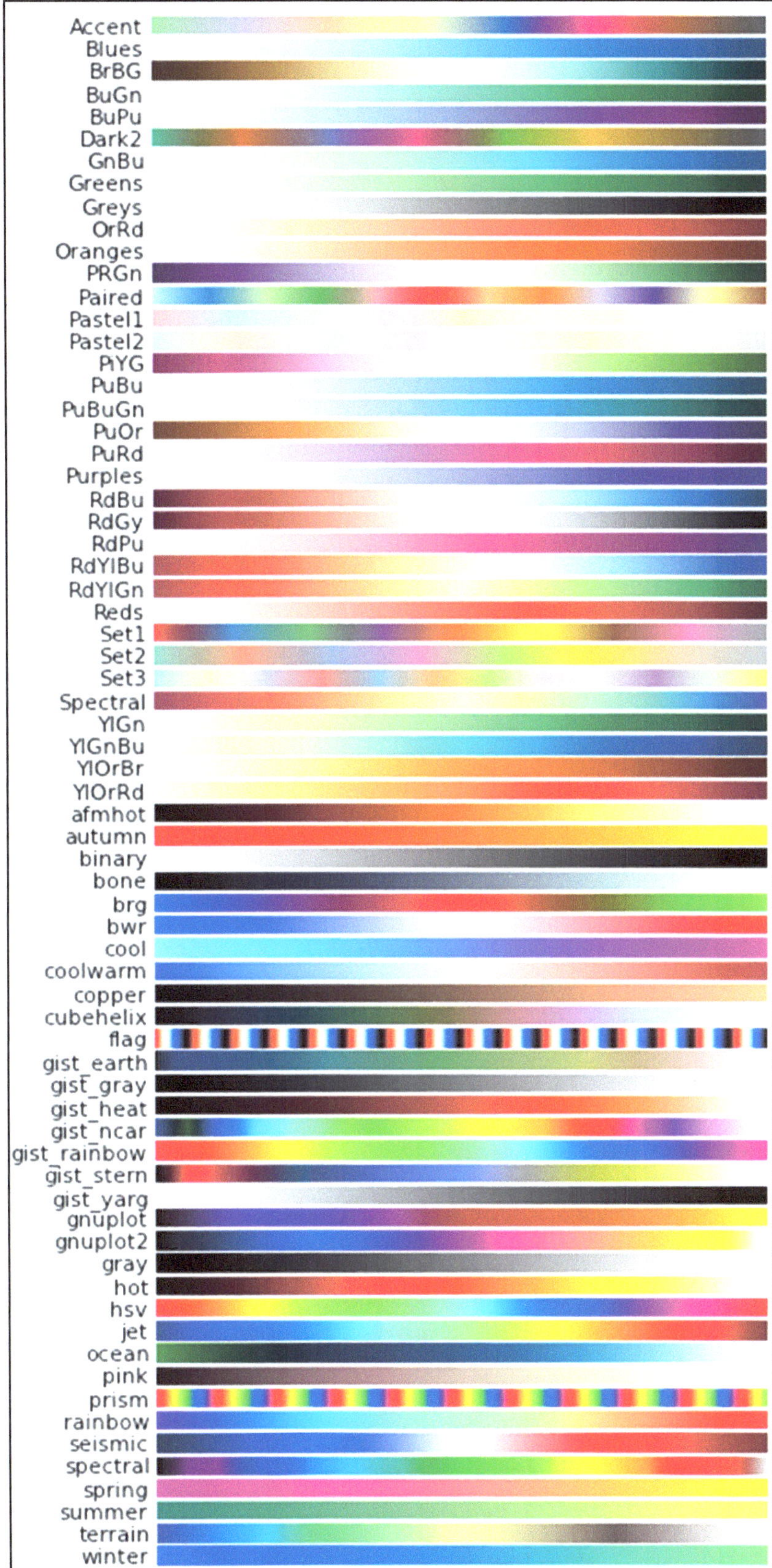

Figure 3-1: Colourmaps in Matplotlib

3.5. MAP MAKING

Now take step ahead with creating a map incorporating Shaded relief map, ETOPO1 relief map, Blue Marble map and thematic map. Import required modules, starting with incorporating *Basemap* class from Basemap toolkit.

#!/usr/bin/env python
from mpl_toolkits.basemap import Basemap

The Matplotlib AxesGrid toolkit is a collection of helper classes, mainly to ease displaying multiple images in Matplotlib. AxesGrid toolkit has been a part of Matplotlib since version 0.99. Originally, the toolkit had a single namespace *axes_grid*. In more recent version, the toolkit has been divided into two separate namespaces, namely, *axes_grid1* and *axisartist*. While *axes_grid* namespace is maintained for the backward compatibility, use of *axes_grid1* and *axisartist* is recommended. *axes_grid1* is a collection of helper classes that provides a framework to adjust the position of multiple axes at drawing time. In Matplotlib, the axes location and size is specified in the normalized figure coordinates, which it might not be ideal for displaying images that needs to have a given aspect ratio. *mpl_toolkits.axes_grid.inset_locator* provides helper classes and functions to place axes (inset) at the anchored position of the parent axes. Using *mpl_toolkits.axes_grid.inset_locator.inset_axes()*, one can have inset axes whose size is either fixed, or a fixed proportion of the parent axes.

from mpl_toolkits.axes_grid1.inset_locator import inset_axes

Import Pyplot module which provide MATLAB-like plotting framework.

import matplotlib.pyplot as plt

Import Polygon class for creating polygon patch.

from matplotlib.patches import Polygon

Import NumPy module for fast array operations.

import numpy as np

The *figure()* function create a new figure and return a *matplotlib.figure.Figure* class instance. Also specify the parameters, *figsize* (width x height in inches) and *facecolor* (background colour).

fig=plt.figure(figsize=(15,15),facecolor='0.8')

Assign a title to the figure using *set_window_title()* method.

fig.canvas.set_window_title(title="Cartography")

Add a centred title to the figure. The parameter *t* accepts the figure title; while *x* and *y* parameters takes the location of text in figure coordinates, *color* parameter accepts any Matplotlib supported colour, *fontsize* can be relative value of *xx-small, x-small, small, medium, large, x-large, xx-large* or an absolute font size, e.g. 12; *fontweight* can be a numeric value in the range 0-1000 or one of *ultralight, light, normal, regular, book, medium, roman, semibold, demibold, demi, bold, heavy, extra bold, black,* the weights are given in table 3-5; *fontstyle* can be either *normal, italic* or *oblique*.

```
fig.suptitle(t='Maps',x=0.5,y=0.5,color='r',fontsize=22,fontweight='bold', \
        fontstyle='italic')
```

Table 3-5: Numerical values of font weights

Font weight	Value
ultralight	100
light	200
normal	400
regular	400
book	400
medium	500
roman	500
semibold	600
demibold	600
demi	600
bold	700
heavy	800
extra bold	800
black	900

Add four subplots in form of two rows and two columns, making the first subplot as current.

```
ax1=fig.add_subplot(221)
```

The above command can be re-written as:

```
ax1=fig.add_subplot(2,2,1)
```

Initiate a *Basemap* class instance with transverse mercator map projection. The desired projection is set using *projection* keyword (default is *cyl*). For most map projections, the map projection region can either be specified by setting the keywords given in table 3-6:

Table 3-6: Keywords for setting map projection region

Keyword	Description
llcrnrlon	longitude of lower left hand corner of the desired map domain (degrees).
llcrnrlat	latitude of lower left hand corner of the desired map domain (degrees).
urcrnrlon	longitude of upper right hand corner of the desired map domain (degrees).

Keyword	Description
urcrnrlat	latitude of upper right hand corner of the desired map domain (degrees).

OR

Keyword	Description
width	width of desired map domain in projection coordinates (metre).
height	height of desired map domain in projection coordinates (metre).
lon_0	centre of desired map domain (in degrees).
lat_0	centre of desired map domain (in degrees).

```
bmap1=Basemap(projection='tmerc',lon_0=82,lat_0=22,k_0=0.9996, \
        rsphere=(6378137.00,6356752.314245179),resolution='l', \
        width=3000000,height=3500000,ax=ax1)
```

Resolution parameter is used to define resolution of boundary database, which can take *c (crude)*, *l (low)*, *I (intermediate)*, *h (high)*, *f (full)* or *None* (default as *c*). If *None*, no boundary dataset is associated with the *Basemap* instance. Higher resolution datasets are much slower to draw. Coastline data is acquired from GSHHS (*http://www.soest.hawaii.edu/wessel/gshhs/*); while state, country and river datasets from the Generic Mapping Tools (*http://gmt.soest.hawaii.edu*). The keyword *ax* sets default axes instance. *k_0* keyword defines scale factor at natural origin (used by *tmerc*, *omerc*, *stere* and *lcc*). *rsphere* parameter takes radius of the sphere used to define map projection (default 6370997 metre, close to the arithmetic mean radius of the earth). If given as a sequence, the first two elements are interpreted as the radii of the major and minor axes of an ellipsoid.

Display ETOPO1 relief image (acquired from *http://www.ngdc.noaa.gov/mgg/global/global.html*) using *etopo()* method, as map background. The default image size is 5400x2700, which can be quite slow and memory exhaustive. The *scale* parameter can be used to downsample the image (*scale=0.5* downsamples to 5400x2700).

```
bmap1.etopo(scale=0.5,ax=ax1)
```

Set the axes title using *set_title()* method.

```
ax1.set_title(label="ETOPO Relief Map",color='k',fontsize=14)
```

Make the second subplot as current.

```
ax2=fig.add_subplot(222)
```

The subplot will be a thematic map of America, depicting land (including state boundaries) and water bodies by different colours. For this, create an instance of *Basemap* class having transverse mercator projection system.

```
bmap2=Basemap(projection='tmerc',lon_0=-90,lat_0=40,k_0=0.9996, \
        rsphere=(6378137.00,6356752.314245179),resolution='c', \
```

width=6000000,height=6000000,ax=ax2)

Draw boundary around projected map region using *drawmapboundary()* method, optionally filling interior of region. The *linewidth* parameter defines line width for boundary (default 1), *color* defines colour of boundary line (default *black*), *fill_color* defines the colour to be filled in the map background (default is to fill with axis background colour); if set to *None*, no filling will be done.

bmap2.drawmapboundary(color='k',linewidth=1.0,fill_color='aqua',ax=ax2)

Draw coastlines using *drawcoastlines()* method. The *linewidth* parameter defines coastline width (default 1), *color* defines coastline colour (default *black*), *antialiased* is the antialiasing switch for coastlines (default *True*).

bmap2.drawcoastlines(linewidth=1.0,color='k',antialiased=1,ax=ax2)

Fill colour in the continent region of the map using *fillcontinents()* method. The *color* parameter defines the colour to fill the continents (default *gray*), while *lake_color* parameter is used to assign colour to fill the inland lakes (default axes background).

bmap2.fillcontinents(color='coral',lake_color='aqua',ax=ax2)

Country boundaries are drawn using *drawcountries()* method. Country boundary line width is defined using *linewidth* (default 0.5), *color* parameter defines country boundary line colour (default *black*), while antialiasing switch for country boundaries is handled by *antialiased* (default *True*).

bmap2.drawcountries(linewidth=1.0,color='k',antialiased=1,ax=ax2)

Draw state boundaries in America *drawstates()* method; *linewidth* parameter controls state boundary line width (default 0.5), *color* parameter defines state boundary line colour (default *black*), and *antialiased* parameter is the antialiasing switch for state boundaries (default *True*).

bmap2.drawstates(linewidth=0.5,color='g',antialiased=1,ax=ax2)

Set the axes title using *set_title()* method.

ax2.set_title(label="Thematic Map",color='k',fontsize=14)

As mentioned earlier, one can have inset axes whose size is either fixed, or a fixed proportion of the parent axes using *mpl_toolkits.axes_grid.inset_locator.inset_axes()* method. The parameter *loc* takes the location of inset axes and it can take string or numeric values as given in table 3-7.

Table 3-7: Location codes

Location	Code
best	0

Location	Code
upper right	1
upper left	2
lower left	3
lower right	4
right	5
center left	6
center right	7
lower center	8
upper center	9
center	10

axin=inset_axes(parent_axes=bmap2.ax,width="30%",height="30%",loc=4)

Define a *Basemap* class instance, followed by drawing countries and filling continent with colour.

omap=Basemap(projection='ortho',lon_0=-105,lat_0=40,ax=axin)
omap.drawcountries(color='white',ax=axin)
omap.fillcontinents(color='gray',ax=axin)

Using *boundarylats/boundarylons* fields (NumPy arrays describing map boundaries) of *Basemap* class, get extent coordinates of inset map boundary.

bx,by=omap(bmap2.boundarylons,bmap2.boundarylats)

Arrange data using *zip()* method, which returns a list of tuples, where the ith tuple contains the ith element from each of the argument sequences or iterables. The returned list is truncated in length to the length of the shortest argument sequence.

xy=zip(bx,by)

Create and add a polygon patch.

mapboundary=Polygon(xy,edgecolor='red',linewidth=2,fill=False)
omap.ax.add_patch(mapboundary)

Make the third subplot as current.

ax3=fig.add_subplot(223)

Let the third subplot be a shaded relief map. For this, define a *Basemap* class instance having transverse mercator projection system.

*bmap3=Basemap(projection='tmerc',lon_0=82,lat_0=22,k_0=0.9996, *

```
rsphere=(6378137.00,6356752.314245179),resolution=None, \
width=3000000,height=3500000,ax=ax3)
```

Display shaded relief image (acquired from *http://www.shadedrelief.com*) as map background using *shadedrelief()* method. The default image size is 10800x5400, which can be quite slow and use quite a bit of memory. The *scale* parameter can be used to downsample the image (*scale=0.5* downsamples to 5400x2700).

```
bmap3.shadedrelief(scale=0.35,ax=ax3)
```

Set variables for map scale placement location based on map region extent.

```
x1,y1=0.75*bmap3.xmax,0.1*bmap3.ymax
```

Calling a *Basemap* class instance with the arguments longitude/latitude (in degrees) will convert to X, Y map projection coordinates (in metre). The inverse transformation is done if the parameter *inverse* is set to *True* (default is *False*). Input arguments longitude/latitude can be scalar floats, sequences or NumPy arrays.

```
lon1,lat1=bmap3(x1,y1,inverse=True)
```

drawmapscale() method draw a map scale at *lon,lat* of length *length* representing distance in the map projection coordinates at *lon0, lat0*. *units* keyword defines the units of the length argument (Default *km*). *barstyle* can be *simple* or *fancy* (default *simple*). *fontsize* (default 9) and *color* (default *black*) are for map scale annotations. *labelstype* can be *simple* (default) or *fancy*; for *fancy*, the map scale factor (ratio between the actual distance and map projection distance at *lon0, lat0*) and the value of *lon0,lat0* are also displayed on the top of the scale bar, and for *simple*, just the units are display on top and the distance below the scale bar; if assigned as *False*, no label will be plotted. *format* is a string formatter to format numeric values. *yoffset* controls how tall the scale bar is, and how far the annotations are offset from the scale bar; default is 0.02 times the height of the map (*0.02*(ymax-ymin)*). *fillcolor1* and *fillcolor2* defines colours of the alternating filled regions (default *white* and *black*); only relevant for *fancy* barstyle.

```
bmap3.drawmapscale(lon=lon1,lat=lat1,lon0=lon1,lat0=lat1,length=1000, \
        barstyle='fancy',units='km',fontsize=9,yoffset=None, \
        labelstyle='simple',fontcolor='k',fillcolor1='w', \
        fillcolor2='k',format='%d',ax=ax3)
```

Set the axes title using *set_title()* method.

```
ax3.set_title(label="Shaded Relief Map",color='k',fontsize=14)
```

Make the fourth subplot as current.

```
ax4=fig.add_subplot(224)
```

Now prepare a map having Blue Marble image as background and drawing a greatcircle arc between New Delhi and Shanghai cities. For this, define a *Basemap* class instance having transverse mercator projection system.

```
bmap4=Basemap(projection='tmerc',lon_0=95,lat_0=22,k_0=0.9996, \
        rsphere=(6378137.00,6356752.314245179),resolution='c', \
        width=9500000,height=7500000,ax=ax4)
```

Prepare an NumPy array having meridian (longitude lines) of 10 degree spacing.

```
meridians=np.arange(0,360,10)
```

Draws and label meridians for values (in degrees) given in the sequence *meridians*. *color* and *linewidth* parameters defines meridian colour (default *black*) and line width (default 1), respectively. *dashes* defines dash pattern for meridians (default [1,1], i.e. 1 pixel on, 1 pixel off). *labels* is a list of 4 values (default [0,0,0,0]) that control whether meridians are labelled where they intersect the left, right, top or bottom of the plot. For example, *labels=[1,0,0,1]* will cause meridians to be labelled where they intersect the left and and bottom of the plot, but not the right and top. *labelstyle* (default *None*) if set to '+/-', then east and west longitudes will be labelled with '+' and '-', otherwise they are labelled with 'E' and 'W'. *xoffset* defines the label offset from edge of map in X-direction (default is 0.01 times width of map in map projection coordinates). *yoffset* defines the label offset from edge of map in Y-direction (default is 0.01 times height of map in map projection coordinates). *latmax* assigns the absolute value of latitude to which meridians are drawn (if *None*, then default value is 80).

```
bmap4.drawmeridians(meridians,color='y',linewidth=1.0,dashes=[1,1], \
            labels=[0,0,0,1],labelstyle=None,xoffset=None,yoffset=None, \
            latmax=None,ax=ax4)
```

Prepare an NumPy array having parallel (latitude lines) of 10 degree spacing.

```
circles=np.arange(-90,90,10)
```

Draw and label parallels for values (in degrees) given in the sequence circles using *drawparallels()* method.

```
bmap4.drawparallels(circles,color='y',linewidth=1.0,dashes=[1,1], \
            labels=[1,0,0,0],labelstyle=None,xoffset=None,yoffset=None, \
            latmax=None,ax=ax4)
```

Prepare lists having latitude/longitude of cities, namely, New Delhi and Shanghai.

```
lat=[28.63,31.23]
lon=[77.22,121.47]
cities=['New Delhi','Shanghai']
```

Draws a great circle on the map from the longitude/latitude pair *lon1, lat1* to *lon2, lat2* using *drawgreatcircle()*.

```
bmap4.drawgreatcircle(lon1=lon[0],lat1=lat[0],lon2=lon[1],lat2=lat[1], \
        linewidth=2,color='m',ax=ax4)
```

Plot city names New Delhi and Shanghai on map.

```
x,y=bmap4(lon,lat)
plt.plot(x,y,'ro')
for city,xpt,ypt in zip(cities,x,y):
  plt.text(x=xpt-1000000,y=ypt+200000,s=city,bbox=dict(facecolor='yellow', \
      alpha=0.5))
```

Draw map scale of *simple* barstyle.

```
bmap4.drawmapscale(lon=110,lat=-5,lon0=110,lat0=-5,length=2000, \
        barstyle='simple',units='km',fontsize=9,yoffset=None, \
        labelstyle='simple',fontcolor='r',format='%d',ax=ax4)
```

Displays Blue Marble image (acquired from *http://visibleearth.nasa.gov*) as map background using *bluemarble()*. The default image size is 5400x2700, which can be quite slow and use quite a bit of memory. The *scale* keyword can be used to downsample the image (*scale=0.5* downsamples to 2700x1350).

```
bmap4.bluemarble(scale=0.5,ax=ax4)
```

Set the axes title using *set_title()* method.

```
ax4.set_title(label="Blue Marble Map",color='k',fontsize=14)
```

Save the current figure using *savefig()* method. The *fname* parameter is a string containing path to a filename or a Python file object; *dpi* is the resolution in dots per inch; *facecolor* indicates the colour of the figure rectangle; *bbox_inches* is the bounding box in inches, which is the portion of the figure being saved, if *tight*, the method tries for a tight fit bounding box of the figure; *pad_inches* is the amount of padding around the figure when *bbox_inches* is *tight.*

```
fname='C:/Results/Maps.png'
plt.savefig(fname,dpi=300,facecolor='0.8',bbox_inches='tight',pad_inches=0.5)
```

Assign the class instances to *None*.

```
fig=bmap1=bmap2=bmap3=bmap4=None
```

Figure 3-2 depicts the map which has been generated using the above discussed script.

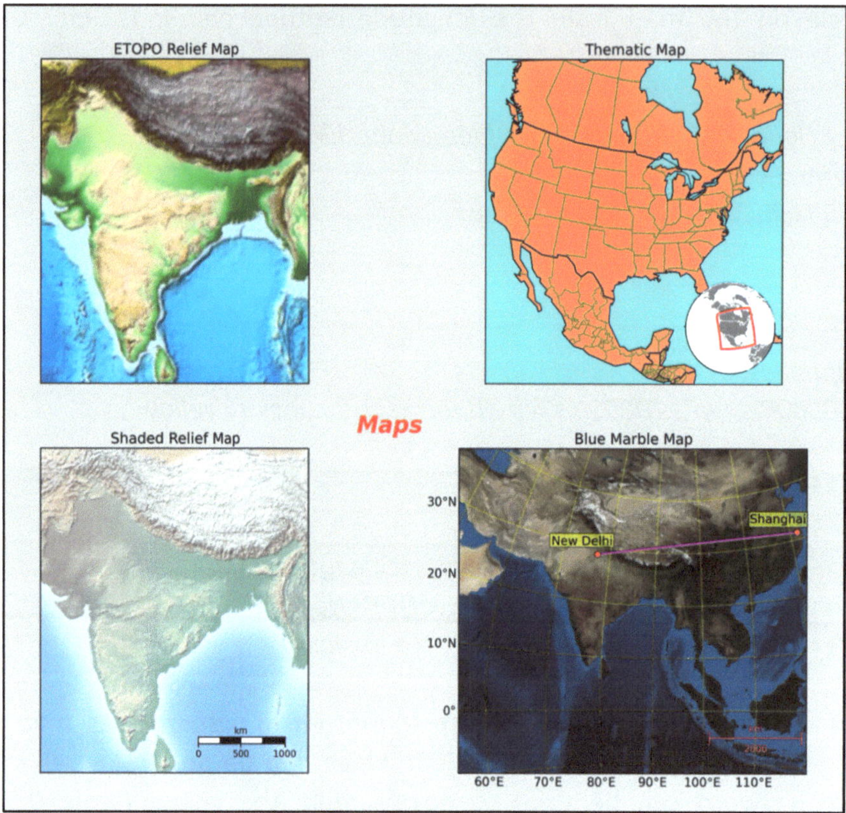

Figure 3-2: Cartography

BIBLIOGRAPHY

- _init_.py from Matplotlib Basemap Toolkit 1.0.5 package.
- axes.py, Matplotlib 1.1.1 package.
- colors.py, Matplotlib 1.1.1 package.
- figure.py, Matplotlib 1.1.1 package.
- font_manager.py, Matplotlib 1.1.1 package.
- legend.py, Matplotlib 1.1.1 package.
- pyplot.py, Matplotlib 1.1.1 package.
- ETOPO1 Global Relief Model, *http://www.ngdc.noaa.gov/mgg/global/global.html* [accessed on 21/05/2012].
- Generic Mapping Tool, *http://gmt.soest.hawaii.edu/* [accessed on 12/05/2012].
- Matplotlib- GitHub Hosting, *http://matplotlib.github.com/* [accessed on 12/06/2012].
- Matplotlib- SourceForge Hosting, *http://matplotlib.sourceforge.net/* [accessed on 30/07/2012].
- Global Self-consistent, Hierarchical, High-resolution Shoreline Database, *http://www.soest.hawaii.edu/wessel/gshhs/* [accessed on 30/04/2012].
- Shaded Relief, *http://www.shadedrelief.com* [accessed on 02/08/2012].
- SourceForge, *http://sourceforge.net/* [accessed on 02/08/2012].
- Visible Earth, *http://visibleearth.nasa.gov* [accessed on 12/08/2012].

Chapter 4
RASTER AND VECTOR DATA PROCESSING

This chapter gives an insight on some basic operations on raster and vector data using GDAL/OGR (version 1.9.1) package. This will be accomplished by discussing scripts for the following tasks:

- Supported raster and vector file formats by GDAL/OGR.
- Extracting information from raster data.
- Creating NDVI.
- Reprojection and resampling.
- Extracting information from shapefile (vector data).
- Raster clipping using shapefile.
- Contour generation.

As the default Python(x,y) setup has GDAL/OGR version 1.9.0, so one can upgrade it by downloading (followed by installing) the latest version of executable file (GDAL/OGR 1.9.1) from link: *http://code.google.com/*. The end of the chapter gives an overview of command-line utility programs of GDAL/OGR.

4.1. GDAL/OGR DRIVERS

Start with a simple program by printing the GDAL/OGR version and its supported file formats. For this, import GDAL and OGR modules which are incorporated in *osgeo* package.

```
#!/usr/bin/env python
try:
    import osgeo.gdal as gdal
    import osgeo.ogr as ogr
except ImportError:
    import gdal
    import ogr
```

Get the GDAL/OGR version information; table 4-1 mentions the possible arguments for *VersionInfo()* methods and their corresponding output.

Table 4-1: GDAL/OGR information

Argument	Output
"VERSION_NUM"	1910
"RELEASE_DATE"	20120516
"RELEASE_NAME"	1.9.1
"--version"	GDAL 1.9.1, released 2012/05/16
"LICENSE"	Returns the license information of GDAL/OGR

```
print gdal.VersionInfo("--version")
```

GDAL/OGR supports reading of many file formats, while it entertains writing of few formats. Fetch the number of registered GDAL drivers. A driver is an object that knows how to interact with a certain type of data. There is a driver for each supported format and one needs to register it before opening a GDAL supported raster data.

```
gdal_driver_count=gdal.GetDriverCount()
print "\nNumber of GDAL drivers: ",gdal_driver_count
```

Fetch all GDAL driver one by one using *GetDriver()*. The *GetDescription()* method fetches the object description; in case of driver object, its short name is returned. Use *GetMetadataItem()* method to get long name of the driver. Table 4-2 provides a comprehensive list of raster data formats that are supported by GDAL.

```
print "\nGDAL drivers (Short name:Long name)"
for i in range(gdal_driver_count):
  driver=gdal.GetDriver(i)
  print driver.GetDescription()+'\t\t: '+driver.GetMetadataItem('DMD_LONGNAME')
```

Table 4-2: GDAL supported raster file formats

Short name/code	Long Format Name
AAIGrid	Arc/Info ASCII Grid
ACE2	ACE2
ADRG	ARC Digitized Raster Graphics
AIG	Arc/Info Binary Grid
AirSAR	AirSAR Polarimetric Image
BIGGIF	Graphics Interchange Format (.gif)
BLX	Magellan topo (.blx)
BMP	MS Windows Device Independent Bitmap
BSB	Maptech BSB Nautical Charts
BT	VTP .bt (Binary Terrain) 1.3 Format
CEOS	CEOS Image
COASP	DRDC COASP SAR Processor Raster
COSAR	COSAR Annotated Binary Matrix (TerraSAR-X)
CPG	Convair PolGASP
CTG	USGS LULC Composite Theme Grid
DIMAP	SPOT DIMAP
DIPEx	DIPEx
DOQ1	USGS DOQ (Old Style)
DOQ2	USGS DOQ (New Style)
DTED	DTED Elevation Raster
E00GRID	Arc/Info Export E00 GRID
ECRGTOC	ECRG TOC format
EHdr	ESRI .hdr Labelled

Short name/code	Long Format Name
EIR	Erdas Imagine Raw
ELAS	ELAS
ENVI	ENVI .hdr Labelled
ERS	ERMapper .ers Labelled
ESAT	Envisat Image Format
FAST	EOSAT FAST Format
FIT	FIT Image
FujiBAS	Fuji BAS Scanner Image
GenBin	Generic Binary (.hdr Labelled)
GFF	Ground-based SAR Applications Testbed File Format (.gff)
GIF	Graphics Interchange Format (.gif)
GRASSASCIIGrid	GRASS ASCII Grid
GRIB	GRIdded Binary (.grb)
GS7BG	Golden Software 7 Binary Grid (.grd)
GSAG	Golden Software ASCII Grid (.grd)
GSBG	Golden Software Binary Grid (.grd)
GSC	GSC Geogrid
GTiff	GeoTIFF
GTX	NOAA Vertical Datum .GTX
GXF	GeoSoft Grid Exchange Format
HF2	HF2/HFZ heightfield raster
HFA	Erdas Imagine Images (.img)
HTTP	HTTP Fetching Wrapper
IDA	Image Data and Analysis
ILWIS	ILWIS Raster Map
INGR	Intergraph Raster
ISIS2	USGS Astrogeology ISIS cube (Version 2)
ISIS3	USGS Astrogeology ISIS cube (Version 3)
JAXAPALSAR	JAXA PALSAR Product Reader (Level 1.1/1.5)
JDEM	Japanese DEM (.mem)
JP2OpenJPEG	JPEG-2000 driver based on OpenJPEG library
JPEG	JPEG JFIF
KMLSUPEROVERLAY	Kml Super Overlay
L1B	NOAA Polar Orbiter Level 1b Data Set
LAN	Erdas .LAN/.GIS
LCP	FARSITE v.4 Landscape File (.lcp)
Leveller	Leveller heightfield
LOSLAS	NADCON .los/.las Datum Grid Shift
MEM	In Memory Raster
MFF	Vexcel MFF Raster
MFF2	Vexcel MFF2 (HKV) Raster

Short name/code	Long Format Name
MSGN	EUMETSAT Archive native (.nat)
NDF	NLAPS Data Format
NGSGEOID	NOAA NGS Geoid Height Grids
NITF	National Imagery Transmission Format
NTv2	NTv2 Datum Grid Shift
NWT_GRC	Northwood Classified Grid Format .grc/.tab
NWT_GRD	Northwood Numeric Grid Format .grd/.tab
OZI	OziExplorer Image File
PAux	PCI .aux Labelled
PCIDSK	PCIDSK Database File
PCRaster	PCRaster Raster File
PDF	Geospatial PDF
PDS	NASA Planetary Data System
PNG	Portable Network Graphics
PNM	Portable Pixmap Format (netpbm)
PostGISRaster	PostGIS Raster driver
R	R Object Data Store
Rasterlite	Rasterlite
RIK	Swedish Grid RIK (.rik)
RMF	Raster Matrix Format
RPFTOC	Raster Product Format TOC format
RS2	RadarSat 2 XML Product
RST	Idrisi Raster A.1
SAGA	SAGA GIS Binary Grid (.sdat)
SAR_CEOS	CEOS SAR Image
SDTS	SDTS Raster
SGI	SGI Image File Format 1.0
SNODAS	Snow Data Assimilation System
SRP	Standard Raster Product (ASRP/USRP)
SRTMHGT	SRTMHGT File Format
Terragen	Terragen heightfield
TIL	EarthWatch .TIL
TSX	TerraSAR-X Product
USGSDEM	USGS Optional ASCII DEM (and CDED)
VRT	Virtual Raster
WCS	OGC Web Coverage Service
WMS	OGC Web Map Service
XPM	X11 PixMap Format
XYZ	ASCII Gridded XYZ
ZMap	ZMap Plus Grid

Get the number of registered OGR drivers.

```
ogr_driver_count=ogr.GetDriverCount()
print "\nNumber of OGR drivers: ",ogr_driver_count
```

Fetch all OGR drivers one by one using *GetDriver()*. The *GetName()* method fetch the name of OGR driver (file format). This name is relatively short (10-40 characters), and reflect the underlying file format. A comprehensive list of vector data formats that are supported by OGR is given in table 4-3.

```
print "\nOGR drivers (Short name)"
for i in range(ogr_driver_count):
   driver=ogr.GetDriver(i)
   print driver.GetName()
```

Table 4-3: OGR supported vector file formats

Short name/code	Format Name
AeronavFAA	Aeronav FAA files
ARCGEN	Arc/Info Generate
AVCBin	Arc/Info Binary Coverage
AVCE00	Arc/Info .E00 (ASCII) Coverage
BNA	Atlas BNA
CouchDB	CouchDB / GeoCouch
CSV	Comma Separated Value (.csv)
DGN	Microstation DGN
DXF	AutoCAD DXF
EDIGEO	EDIGEO
ESRI Shapefile	ESRI Shapefile
Geoconcept	Géoconcept Export
GeoJSON	GeoJSON
Geomedia	Geomedia .mdb
GeoRSS	GeoRSS
GFT	Google Fusion Tables
GML	GML
GMT	GMT
GPSBabel	GPSBabel
GPSTrackMaker	GPSTrackMaker (.gtm, .gtz)
GPX	GPX
HTF	Hydrographic Transfer Format
Idrisi	Idrisi Vector (.VCT)
KML	KML
MapInfo File	Mapinfo File
Memory	Memory
MSSQLSpatial	MS SQL Spatial
MySQL	MySQL
NAS	NAS - ALKIS
ODBC	ODBC
OpenAir	OpenAir

Short name/code	Format Name
PCIDSK	PCI Geomatics Database File
PDS	PDS
PGDump	PGDump
PGeo	ESRI Personal GeoDatabase
PostgreSQL	PostgreSQL
REC	EPIInfo .REC
S57	S-57 (ENC)
SDTS	SDTS
SEGUKOOA	SEG-P1 / UKOOA P1/90
SEGY	SEG-Y
SQLite	SQLite/SpatiaLite
SUA	SUA
SVG	SVG
TIGER	U.S. Census TIGER/Line
UK .NTF	UK .NTF
VFK	VFK data
VRT	VRT - Virtual Datasource
WFS	OGC WFS (Web Feature Service)
XPlane	X-Plane/Flighgear aeronautical data

4.2. RASTER INFORMATION

Now take a step further and try to write a script which can be used to extract some meaningful information from raster data. Initiate with importing required modules.

```
#!/usr/bin/env python
try:
   import osgeo.gdal as gdal
   import osgeo.gdal_array as gdal_array
   import osgeo.osr as osr
except ImportError:
   import gdal
   import gdal_array
   import osr
import matplotlib.pyplot as plt
import numpy as np
import sys,os
```

Register all GDAL drivers; Python automatically calls *gdal.AllRegister()* when GDAL module is imported. Generally, this method is called once at the beginning of the application.

```
gdal.AllRegister()
```

Let the test data be *small_world.tiff*, which is three bands GeoTIFF imagery. Assign the complete path of the file to a variable and then call *Open()* method to open the dataset by passing the name of dataset and the desired access (default is *GA_ReadOnly*). Also note that, the dataset passed to *Open()* method need not actually be the name of a physical file (though it usually is). Its interpretation is driver dependent, and it might be a URL or a filename. Many drivers support only read only access, but the available data access options are given in table 4-4:

Table 4-4: Dataset access options

Access code		Description
Numeric	**String**	
0	GA_ReadOnly	Read only (no update) access
1	GA_Update	Read/write access.

```
file='C:/Data/small_world.tiff'
dataset=gdal.Open(file,gdal.GA_ReadOnly)
```

A dataset is an assembly of related raster bands which has some common information, such as, georeferencing transform and coordinate system definition. The dataset can have metadata in the form of name and/or values in string format.

It is always a good idea to check and make sure that the file is actually opened or not. The *Open()* method returns the dataset on successful opening using driver, while it returns *None* on failure. If the opening of file fails, print an error message and terminate the script.

```
if dataset is None:
   print 'Unable to open file'
   sys.exit(1)
```

After getting the dataset object, the actual data need to be extracted out of it. The simplest approach is to use *DatasetReadAsArray()* method, which makes an entire copy of the dataset (as NumPy array), unless the data are explicitly subsetted as part of the function call. For large data, this approach is expected to be prohibitively memory intensive.

```
AllRasterData=gdal_array.DatasetReadAsArray(ds=dataset,xoff=0,yoff=0, \
                      xsize=None,ysize=None)
```

In the above method, the parameter *ds* takes the dataset as argument, *xoff* and *yoff* takes the offset values to start reading at (default arguments of 0), while *xsize* and *ysize* accepts number of pixels to read in X and Y direction (default arguments of 0).

Fetch the driver which was used to open the file and print its short and long name. For example, the short name of GeoTIFF driver is *GTiff* while its long name is *GeoTIFF*.

```
print("Driver: %s/%s" %(dataset.GetDriver().ShortName, \
```

```
            dataset.GetDriver().LongName))
```

Fetch files forming dataset using *GetFileList()*. This method returns a list of files, believed to be part of the dataset. If it returns an empty list of files, it is believed that there are no files associated with the dataset (for instance, a virtual dataset). The returned filenames (as UTF-8 encoded strings) will normally be relative or absolute paths depending on the path used to originally open the dataset.

```
FileList=dataset.GetFileList()
if FileList is None or len(FileList)==0:
  print("Files: none associated")
else:
  print("Files: %s" %FileList[0])
  for i in range(1, len(FileList)):
    print("      %s" %FileList[i])
```

Get dataset size, i.e. width (columns) and height (rows) of raster band in pixels, and number of bands in dataset.

```
print("Size is %dx%dx%d" %(dataset.RasterYSize,dataset.RasterXSize, \
                        dataset.RasterCount))
```

GetProjectionRef() method returns a string having coordinate system of the dataset in OpenGIS WKT format, else an empty string (‘’) is returned.

```
Projection=dataset.GetProjectionRef()
```

If there is coordinate system string present, define an SRS instance. Then use *ImportFromWkt()* method, which wipes the existing SRS definition, and reassign it based on the contents of the passed WKT string; returning 0 upon failure. Print a nicely formatted WKT string for display using *ExportToPrettyWkt()*; if *True* is passed as an argument, then the method simplifies the SRS by stripping the *AXIS*, *AUTHORITY* and *EXTENSION* nodes.

```
if Projection is not '':
  SRS=osr.SpatialReference()
  if SRS.ImportFromWkt(Projection) is 0:
    PrettyWkt=SRS.ExportToPrettyWkt(False)
    print("Coordinate System is:\n%s" %PrettyWkt)
  else:
    print("Coordinate System is:\n%s" %Projection)
```

GDAL dataset have two ways of describing the relationship between raster positions (in *pixel, line* coordinates) and georeferenced coordinates. The first and most commonly used is the affine transform (the other is GCPs). The affine transform consists of six coefficients returned by *GetGeoTransform()*, which map *pixel, line* coordinates into georeferenced space using equations 4-1a and 4-1b.

$$X_{geo} = GT(0) + X_{pixel} \times GT(1) + Y_{line} \times GT(2) \quad (4\text{-}1a)$$
$$Y_{geo} = GT(3) + X_{pixel} \times GT(4) + Y_{line} \times GT(5) \quad (4\text{-}1b)$$

In a north up image, *GT(1)* is the pixel width, and *GT(5)* is the pixel height; the upper left corner of the upper left pixel is at position (*GT(0), GT(3)*); both *GT(2)* and *GT(4)* are 0, and they are rotation parameters.

```
geotrans=dataset.GetGeoTransform()
if geotrans is not None:
  print 'Geotransformation parameters:'
  print 'Origin(Top left X): ',geotrans[0]
  print 'Origin(Top left Y): ',geotrans[3]
  print 'Pixel size in X direction: ',geotrans[1]
  print 'Pixel size in Y direction: ',geotrans[5]
  print 'Rotation: ',geotrans[2]
  print 'Rotation: ',geotrans[4]
```

A dataset can have a set of control points relating one or more positions on the raster to georeferenced coordinates. All GCPs share a georeferencing coordinate system (returned by *GetGCPProjection()*). *GetGCPCount()* gives number of GCPs for the dataset; returns 0 if there is none. After fetching, extract the attributes of GCPs using *GetGCPs()* method. The *Id* field is a string intended to be a unique (and often, but not always numerical) identifier for the GCP within the set of GCPs of the dataset. The *Info* field is usually an empty string, but can contain any user defined text associated with the GCP. The *GCPPixel, GCPLine* gives the *pixel, line* position of the GCP on the raster. The *GCPX, GCPY, GCPZ* gives the associated georeferenced location (X,Y,Z) with Z often being 0. The GDAL data model does not imply a transformation mechanism that must be generated from the GCPs; this is left to the application. Normally a dataset will contain an affine geotransform, GCPs or neither of two.

```
if dataset.GetGCPCount()>0:
  Projection=dataset.GetGCPProjection()
  if Projection is not None:
    SRS=osr.SpatialReference()
    if SRS.ImportFromWkt(Projection)==0:
      PrettyWkt=SRS.ExportToPrettyWkt(False)
      print("GCP Projection: \n%s" %PrettyWkt)
    else:
      print("GCP Projection: %s" %Projection)
  gcps=dataset.GetGCPs()
  i=0
  for gcp in gcps:
    print("GCP[%3d]: Id=%s, Info=%s\n" \
    "           (%.15g,%.15g)->(%.15g,%.15g,%.15g)" \
 %(i,gcp.Id,gcp.Info,gcp.GCPPixel,gcp.GCPLine, \
                gcp.GCPX,gcp.GCPY,gcp.GCPZ))
```

```
i = i + 1
```

GDAL metadata is auxiliary information, which is application specific textual data, kept as a list of name/value pairs. The names are required to be well behaved tokens (no spaces, or odd characters). The values can be of any length, and contain anything except an embedded null (ASCII zero). Some formats support generic (user defined) metadata, while other format drivers map specific format fields to metadata names. For instance, the TIFF driver returns few information tags as metadata including the date/time field which is returned as:

TIFFTAG_DATETIME=1999:05:11 11:29:56

Metadata is split into named groups called domains, with the default domain having no name (""). Some of the domains will be discussed in this literature. The following metadata items have well defined semantics in the default domain:

- *AREA_OR_POINT*: May be either *Area* (the default) or *Point*. It indicates whether a pixel value should be assumed to represent a sampling over the region of the pixel or a point sample at the centre of the pixel. This is not intended to influence interpretation of georeferencing which remains area oriented.
- *NODATA_VALUES*: The value is a list of space separated pixel values matching the number of bands in the dataset that can be collectively used to identify pixels that are no data in the dataset. With this style of no data, a pixel is considered no data in all bands, if and only if all bands match the corresponding value in the *NODATA_VALUES* tuple. This metadata is not widely honoured by GDAL drivers, algorithms or utilities at this time.
- *MATRIX_REPRESENTATION*: This value, used for Polarimetric SAR datasets, contains the matrix representation that this data is provided in. The following are acceptable values:
    - SCATTERING
    - SYMMETRIZED_SCATTERING
    - COVARIANCE
    - SYMMETRIZED_COVARIANCE
    - COHERENCY
    - SYMMETRIZED_COHERENCY
    - KENNAUGH
    - SYMMETRIZED_KENNAUGH
- *POLARMETRIC_INTERP*: This metadata item is defined for raster bands of polarimetric SAR data. This indicates which entry in the specified matrix representation of the data a given band represents. For dataset provided as a scattering matrix, for example, acceptable values for this metadata item are *HH, HV, VH, VV*. When the dataset is a covariance matrix, for example, this metadata item will be one of *Covariance_11*, *Covariance_22*, *Covariance_33*, *Covariance_12*, *Covariance_13*, *Covariance_23* (since the matrix itself is a hermitian matrix, that is all the data that is required to describe the matrix).

```
Metadata=dataset.GetMetadata_List("")
if Metadata is not None and len(Metadata)>0:
  print("Default domain metadata:")
```

```
for md in Metadata:
    print("%s" %md)
```

The *IMAGE_STRUCTURE* domain of metadata is used to hold structural information about image organization that would not normally be carried with an image when translated into another format. The *IMAGE_STRUCTURE* metadata may occur in the dataset or in individual bands. When items like *NBITS* are found in the dataset, it is assumed that they apply to all bands of that dataset. Currently the following items are defined in the *IMAGE_STRUCTURE* domain:

- *COMPRESSION*: The compression type used for this dataset or band. There is no fixed catalog of compression type names, but where a given format includes a *COMPRESSION* creation option, the same list of values should be used here.
- *NBITS*: The actual number of bits used for a band, or the bands of the dataset. It is normally present when the number of bits is non-standard for the data type, such as when a 1-bit TIFF is represented through GDAL as *GDT_Byte*.
- *INTERLEAVE*: This only applies on datasets, and the value should be one of *PIXEL*, *LINE* or *BAND*. It can be used as a data access hint.
- *PIXELTYPE*: This may appear on a *GDT_Byte* band (or the corresponding dataset) and have the value *SIGNEDBYTE* to indicate the unsigned byte values between 128 and 255 should be interpreted as being values between -128 and -1 for applications that recognise the *SIGNEDBYTE* type.

```
Metadata=dataset.GetMetadata_List("IMAGE_STRUCTURE")
if Metadata is not None and len(Metadata)>0:
  print("IMAGE_STRUCTURE domain metadata:")
  for md in Metadata:
    print("%s" %md)
```

The *SUBDATASETS* domain holds a list of child datasets. Normally this is used to provide pointers to a list of images stored within a single multi image file (such as HDF or NITF). For instance, an NITF with four images might have the following subdataset list.

```
SUBDATASET_1_NAME=NITF_IM:0:multi_1b.ntf
SUBDATASET_1_DESC=Image 1 of multi_1b.ntf
SUBDATASET_2_NAME=NITF_IM:1:multi_1b.ntf
SUBDATASET_2_DESC=Image 2 of multi_1b.ntf
SUBDATASET_3_NAME=NITF_IM:2:multi_1b.ntf
SUBDATASET_3_DESC=Image 3 of multi_1b.ntf
SUBDATASET_4_NAME=NITF_IM:3:multi_1b.ntf
SUBDATASET_4_DESC=Image 4 of multi_1b.ntf
SUBDATASET_5_NAME=NITF_IM:4:multi_1b.ntf
SUBDATASET_5_DESC=Image 5 of multi_1b.ntf
```

The value of the *_NAME* is the string that can be passed to *Open()* method to access the file. The *_DESC* value is intended to be a more user friendly string that can be displayed to the user.

```
Metadata=dataset.GetMetadata_List("SUBDATASETS")
if Metadata is not None and len(Metadata)>0:
  print("SUBDATASETS domain metadata:")
  for md in Metadata:
    print("%s" %md)
```

It is proposed that GDAL should support an additional mechanism for geolocation of imagery (in development stage) based on large arrays of points associating pixels and lines with geolocation coordinates. These arrays would be represented as raster bands themselves. It is common in AVHRR, Envisat, HDF and netCDF data products to distribute geolocation for raw or projected data in this manner, and current approaches of representing this as very large numbers of GCPs, or greatly subsampling the geolocation information to provide more reasonable numbers of GCPs are inadequate for many applications.

Datasets with geolocation information will include the following dataset level metadata items in the *GEOLOCATION* domain to identify the geolocation arrays, and the details of the coordinate system and relationship back to the original pixels and lines.

- *SRS*: wkt encoding of spatial reference system.
- *X_DATASET*: dataset name (defaults to same dataset if not specified).
- *X_BAND*: band number within *X_DATASET*.
- *Y_DATASET*: dataset name (defaults to same dataset if not specified).
- *Y_BAND*: band number within *Y_DATASET*.
- *Z_DATASET*: dataset name (defaults to same dataset if not specified).
- *Z_BAND*: band number within *Z_DATASET* (optional).
- *PIXEL_OFFSET*: pixel offset into geo-located data of left geolocation pixel.
- *LINE_OFFSET*: line offset into geo-located data of top geolocation pixel.
- *PIXEL_STEP*: each geolocation pixel represents this many geolocated pixels.
- *LINE_STEP*: each geolocation pixel represents this many geolocated lines.

```
Metadata=dataset.GetMetadata_List("GEOLOCATION")
if Metadata is not None and len(Metadata)>0:
  print("GEOLOCATION domain metadata:")
  for md in Metadata:
    print("%s" %md)
```

The *RPC* domain metadata holds information describing the Rational Polynomial Coefficient geometry model for the image, if present. This geometry model can be used to transform between *pixel*, *line* and georeferenced locations. The items defining the model are:

- *ERR_BIAS*: Error-Bias. The RMS bias error in metre per horizontal axis of all points in the image (-1.0 if unknown).
- *ERR_RAND*: Error-Random. RMS random error in metre per horizontal axis of each point in the image (-1.0 if unknown).
- *LINE_OFF*: Line Offset.
- *SAMP_OFF*: Sample Offset.

- *LAT_OFF*: Geodetic Latitude Offset.
- *LONG_OFF*: Geodetic Longitude Offset.
- *HEIGHT_OFF*: Geodetic Height Offset.
- *LINE_SCALE*: Line Scale.
- *SAMP_SCALE*: Sample Scale.
- *LAT_SCALE*: Geodetic Latitude Scale.
- *LONG_SCALE*: Geodetic Longitude Scale.
- *HEIGHT_SCALE*: Geodetic Height Scale.
- *LINE_NUM_COEFF* (1-20): Line Numerator Coefficients. Twenty coefficients for the polynomial in the Numerator of the *rn* equation (space separated).
- *LINE_DEN_COEFF* (1-20): Line Denominator Coefficients. Twenty coefficients for the polynomial in the Denominator of the *rn* equation (space separated).
- *SAMP_NUM_COEFF* (1-20): Sample Numerator Coefficients. Twenty coefficients for the polynomial in the Numerator of the *cn* equation (space separated).
- *SAMP_DEN_COEFF* (1-20): Sample Denominator Coefficients. Twenty coefficients for the polynomial in the Denominator of the *cn* equation (space separated).

```
Metadata=dataset.GetMetadata_List("RPC")
if Metadata is not None and len(Metadata)>0:
  print("RPC domain metadata:")
  for md in Metadata:
    print("%s" %md)
```

Geographical extents of raster can be generated using raster size and transformation parameters (using equations 4-1a and 4-1b).

```
print 'Corner coordinates'
print("Upper left: %.5f,%.5f" %(geotrans[0],geotrans[3]))
print("Upper right: %.5f,%.5f" %(geotrans[0]+geotrans[1]*dataset.RasterXSize, \
                  geotrans[3]))
print("Lower left: %.5f,%.5f" %(geotrans[0], \
 geotrans[3]+geotrans[5]*dataset.RasterYSize))
print("Lower right: %.5f,%.5f" %(geotrans[0]+geotrans[1]*dataset.RasterXSize, \
                  geotrans[3]+geotrans[5]*dataset.RasterYSize))
```

Using *for* loop, access raster data on band-by-band (numbered 1 through *RasterCount*; where RasterCount field gives the number of bands in a dataset) basis, and extract metadata, block size, colour table, and other available information. Fetch a band object for dataset using *GetRasterBand()*.

```
for iBand in range(dataset.RasterCount):
  Band=dataset.GetRasterBand(iBand+1)
  print('Band %d information:' %(iBand+1))
```

GDAL contains a concept of the natural block size of raster, so that application can organize data access efficiently for some file formats. The natural block size is the block size that is most efficient for accessing the format. For many formats, this is simple a whole scanline; however, for tiled images this will typically be the tile size. Note that the X and Y block sizes do not have to divide the image size evenly, meaning that right and bottom edge blocks can be incomplete type.

```
BlockXSize,BlockYSize=Band.GetBlockSize()
print("Block size: %dx%d" %(BlockYSize,BlockXSize))
```

Pixel data type of a band can be fetched using *DataType* field, the various data types are given in table 4-5. *GetDataTypeName()* method is used to acquire a symbolic name for the data type. This is essentially the enumerated item name with the *GDT_* prefix removed, so *GDT_Byte* returns *Byte*.

Table 4-5: Data types of raster band

Data type code		Description
Numeric	**String**	
0	GDT_Unknown	Unknown or unspecified type
1	GDT_Byte	Eight bit unsigned integer
2	GDT_UInt16	Sixteen bit unsigned integer
3	GDT_Int16	Sixteen bit signed integer
4	GDT_UInt32	Thirty two bit unsigned integer
5	GDT_Int32	Thirty two bit signed integer
6	GDT_Float32	Thirty two bit floating point
7	GDT_Float64	Sixty four bit floating point
8	GDT_CInt16	Complex Int16
9	GDT_CInt32	Complex Int32
10	GDT_CFloat32	Complex Float32
11	GDT_CFloat64	Complex Float64
12	GDT_TypeCount	Maximum type

```
print("Data type: %s" %(gdal.GetDataTypeName(Band.DataType)))
```

GetDescription() acquires raster band's description (if supported), or else an empty string ('') is returned.

```
if Band.GetDescription() is not None and len(Band.GetDescription())>0:
  print("Description: %s" %Band.GetDescription())
```

GetMinimum() and *GetMaximum()* fetches the minimum and maximum pixel value for the band, respectively, excluding no data pixels. For file formats that do not know these intrinsically, the minimum and maximum supported values for the data type will generally be returned. *ComputeRasterMinMax()* compute the minimum and maximum values of a band; if approximate statistics are sufficient, the *approx_ok* flag can be set to *True* or 1 (default is 0), in which, overviews or a subset of all image tiles may be used in computing the statistics.

```
min=Band.GetMinimum()
max=Band.GetMaximum()
if min is None or max is None:
  line=""
  if min is not None:
    line=line+("Min: %.3f " %min)
  if max is not None:
    line=line+("Max: %.3f " %max)
  MinMax=Band.ComputeRasterMinMax(False)
  line=line+("Computed Min/Max: %.3f,%.3f" \
            %(MinMax[0],MinMax[1]))
  print(line)
```

Checksum() method computes 16-bit (0-65535) checksum from a region of raster band data. Floating point data is converted to 32-bit integer, so decimal portions of such raster data will not affect the checksum. Real and imaginary components of complex bands influence the result. The *xoff* and *yoff* parameters take pixel and line offset values of window to read (default arguments 0); while *xsize* and *ysize* parameters takes pixel and line size of window to read (default arguments *None*).

```
print("Checksum: %d" %Band.Checksum(xoff=0, \
      yoff=0,xsize=None,ysize=None))
```

Fetch the no data value for a band, which is a special marker value used to mark pixels that are not valid data. Such pixels should generally not be displayed, nor contribute to analysis operations.

```
NoData=Band.GetNoDataValue()
if NoData is not None:
  print("NoData value: %.18g" %NoData)
```

Overviews are copies of an image at a lower resolution. The idea is that if the whole image is viewed, the overview of smallest resolution is shown to minimize performance overhead. As a smaller sub-section of the image is zoomed in, the higher resolution overviews are used until finally the actual image is displayed. A band may have zero or more overviews which can be known using *GetOverviewCount()* method. The size (in pixels and lines) of the overview will be different than the underlying raster, but the geographic region covered by overviews is the same as the full resolution band.

```
if Band.GetOverviewCount()>0:
  line="Overviews: "
  for iOverview in range(Band.GetOverviewCount()):
    if iOverview!=0:
      line=line+", "
    Overview=Band.GetOverview(iOverview)
    if Overview is not None:
      line=line+("%dx%d" %(Overview.XSize,Overview.YSize))
```

```
        else:
            line=line+"(null)"
    print(line)
```

There is a *HasArbitraryOverviews()* method, which returns *True*, if the raster can be read at any resolution efficiently, but with no distinct overview levels.

```
if Band.HasArbitraryOverviews():
    print("Overviews: arbitrary")
```

GetUnitType() method returns a name for the units of the raster values. For instance, it might be *m* for an elevation model in metre, or *ft* for feet. If no unit is available, an empty string ('') will be returned.

```
if len(Band.GetUnitType())>0:
    print("Unit Type: %s" %Band.GetUnitType())
```

GetOffset() method fetches the raster value offset. For file formats that do not know this intrinsically, a value of zero is returned. *GetScale()* fetches the raster value scale. For file formats that do not know this intrinsically, a value of one is returned. This value (in combination with the offset) is used to transform raw pixel values into the units returned by *GetUnitType()*,the mathematical relation is depicted by equation 4-2.

$$\text{Units value}=(\text{raw value x scale})+\text{offset} \tag{4-2}$$

```
if Band.GetScale()!=1.0 or Band.GetOffset()!=0.0:
    print("Offset: %.15g, Scale: %.15g" % \
        (Band.GetOffset(),Band.GetScale()))
```

Get the default and image structure domain metadata of individual bands.

```
Metadata=Band.GetMetadata_List("")
if Metadata is not None and len(Metadata)>0:
    print("Default domain metadata:")
    for md in Metadata:
        print("%s" %md)
Metadata=Band.GetMetadata_List("IMAGE_STRUCTURE")
if Metadata is not None and len(Metadata)>0:
    print("IMAGE_STRUCTURE domain metadata:")
    for md in Metadata:
        print("%s" %md)
```

GetPaletteInterpretationName() returns a symbolic name for the colour palette interpretation. This is the the enumerated item name with the *GPI_* prefix removed. So *GPI_Gray* returns *Gray*. *GetRasterColorInterpretation()* returns the type of colour interpretation of raster band, which is one of the following from the table 4-6. When the format does not know anything about the colour interpretation, *GCI_Undefined* is returned.

Table 4-6: Colour interpretation of raster band

Colour interpretation code		Description
Numeric	String	
0	GCI_Undefined	
1	GCI_GrayIndex	Gray scale
2	GCI_PaletteIndex	Paletted (see associated colour table)
3	GCI_RedBand	Red band of RGBA image
4	GCI_GreenBand	Green band of RGBA image
5	GCI_BlueBand	Blue band of RGBA image
6	GCI_AlphaBand	Alpha (0=transparent, 255=opaque)
7	GCI_HueBand	Hue band of HLS image
8	GCI_SaturationBand	Saturation band of HLS image
9	GCI_LightnessBand	Lightness band of HLS image
10	GCI_CyanBand	Cyan band of CMYK image
11	GCI_MagentaBand	Magenta band of CMYK image
12	GCI_YellowBand	Yellow band of CMYK image
13	GCI_BlackBand	Black band of CMLY image
14	GCI_YCbCr_YBand	Y Luminance
15	GCI_YCbCr_CbBand	Cb Chroma
16	GCI_YCbCr_CrBand	Cr Chroma

GetColorInterpretationName() method returns a symbolic name for the colour interpretation. This is derived from the enumerated item name with the *GCI_* prefix removed, but there are some variations. So, *GCI_GrayIndex* returns *Gray* and *GCI_RedBand* returns *Red*.

```
print("Colour interpretation: %s" \
%(gdal.GetColorInterpretationName(Band.GetRasterColorInterpretation())))
```

A colour table consists of zero or more colour entries and also has a palette interpretation value (fetched by *GetPaletteInterpretation()*), which is one of the following values given in table 4-7, and it indicates how the c1/c2/c3/c4 values of a colour entry should be interpreted.

Table 4-7: Colour palette

Colour palette code		Description
Numeric	String	
0	GPI_Gray	Use c1 as grayscale value
1	GPI_RGB	Use c1 as red, c2 as green, c3 as blue and c4 as alpha
2	GPI_CMYK	Use c1 as cyan, c2 as magenta, c3 as yellow and c4 as black
3	GPI_HLS	Use c1 as hue, c2 as lightness, and c3 as saturation

Fetch the colour table associated with a band using *GetRasterColorTable()* method, and then extract its colour entries.

```
CTable=Band.GetRasterColorTable()
if Band.GetRasterColorInterpretation()==gdal.GCI_PaletteIndex \
                    and CTable is not None:
  print("Colour Table (%s with %d entries)" % \
  (gdal.GetPaletteInterpretationName( \
  CTable.GetPaletteInterpretation()), \
  CTable.GetCount()))
  for i in range(CTable.GetCount()):
    Entry=CTable.GetColorEntry(i)
    print( " %3d: %d,%d,%d,%d" % ( \
    i,Entry[0],Entry[1],Entry[2],Entry[3]))
```

As the raster array is of shape *bands x rows x columns*, one needs to transform it to *rows x columns x bands* for plotting the image using *imshow()*. This task is accomplished using *rollaxis()* method. The plotted image is shown in figure 4-1, while figure 4-2 shows the IPython window having raster information.

```
AllRasterData=np.rollaxis(AllRasterData,0,3)
plt.imshow(X=AllRasterData)
plt.axis("off")
```

Let the plot title be the filename.

```
(filepath,filename)=os.path.split(file)
plt.title(filename)
```

Destroy dataset object by assigning it *None*.

```
dataset=None
```

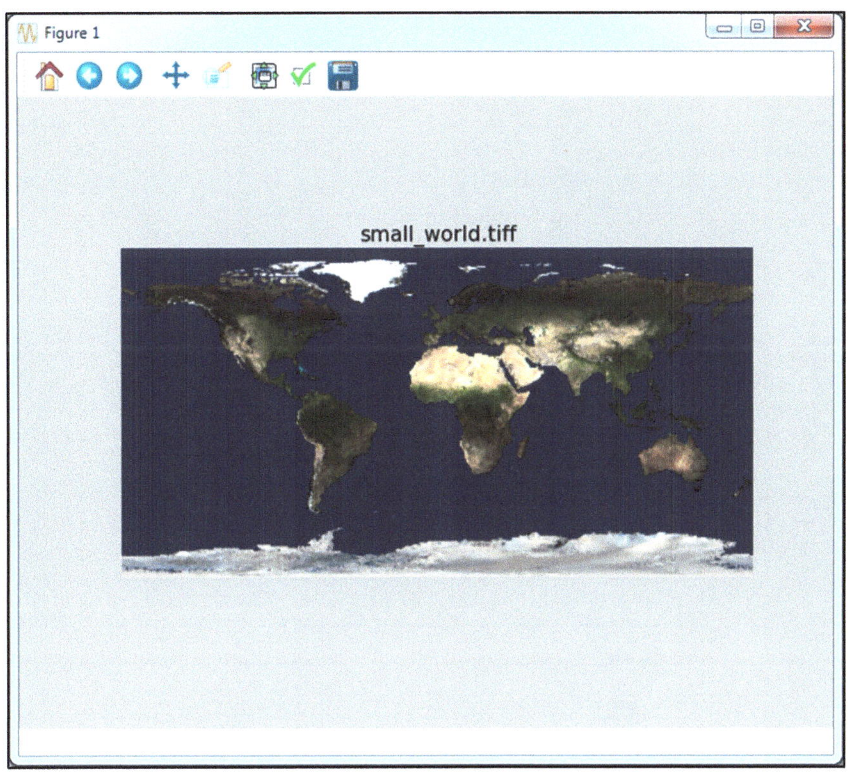

Figure 4-1: Multi-spectral imagery (small_world.tiff) plot

```
Console
IPython 1                                              00:00:16
?         -> Introduction and overview of IPython's features.
%quickref -> Quick reference.
help      -> Python's own help system.
object?   -> Details about 'object'. ?object also works, ?? prints more.

  Welcome to pylab, a matplotlib-based Python environment.
  For more information, type 'help(pylab)'.

In [1]: runfile(r'C:\Scripts\Raster_information.py', wdir=r'C:\Scripts')
Driver: GTiff/GeoTIFF
Files: C:/Data/small_world.tiff
Size is 200x400x3
Coordinate System is:
GEOGCS["WGS 84",
    DATUM["WGS_1984",
        SPHEROID["WGS 84",6378137,298.257223563,
            AUTHORITY["EPSG","7030"]],
        AUTHORITY["EPSG","6326"]],
    PRIMEM["Greenwich",0],
    UNIT["degree",0.0174532925199433],
    AUTHORITY["EPSG","4326"]]
Geotransformation parameters:
Origin(Top left X):  -180.0
Origin(Top left Y):  90.0
Pixel size in X direction:  0.9
Pixel size in Y direction:  -0.9
Rotation:  0.0
Rotation:  0.0
Default domain metadata:
AREA_OR_POINT=Area
IMAGE_STRUCTURE domain metadata:
INTERLEAVE=BAND
Corner coordinates
Upper left: -180.00000,90.00000
Upper right: 180.00000,90.00000
Lower left: -180.00000,-90.00000
Lower right: 180.00000,-90.00000
Band 1 information:
Block size: 20x400
Data type: Byte
Computed Min/Max: 0.000,255.000
Checksum: 30111
Color interpretation: Red
Band 2 information:
Block size: 20x400
Data type: Byte
Computed Min/Max: 0.000,255.000
Checksum: 32302
Color interpretation: Green
Band 3 information:
Block size: 20x400
Data type: Byte
Computed Min/Max: 0.000,255.000
Checksum: 40026
Color interpretation: Blue

In [2]:
```

Figure 4-2: Raster (small_world.tiff) information

4.3. NDVI PREPARATION

Now move to the next level by writing a script for processing extracted raster data by preparing *Normalized Difference Vegetation Index* (NDVI) raster. The script has been tested using LANDSAT imageries (details of test data is mentioned in section 1.12). Start with importing required modules followed by registering all drivers.

```
#!/usr/bin/env python
try:
  import osgeo.gdal as gdal
  import osgeo.gdalconst as gdalconst
except ImportError:
  import gdal
  import gdalconst
import sys,numpy
gdal.AllRegister()
```

Open the red band (band 3 of LANDSAT data) followed by NIR band (band 4), and check if the files are properly opened or not.

```
File_R_Band='C:/Data/p145r045_7t20011020_z43_nn30.tif'
Dataset_R_Band=gdal.Open(File_R_Band,gdalconst.GA_ReadOnly)
File_NIR_Band='C:/Data/p145r045_7t20011020_z43_nn40.tif'
Dataset_NIR_Band=gdal.Open(File_NIR_Band,gdalconst.GA_ReadOnly)
if Dataset_NIR_Band is None or Dataset_R_Band is None:
  print 'Unable to open file'
  sys.exit(1)
```

Obtain the raster sizes of both bands, which is needed for creation of raster for NDVI. Also extract the no data value of both bands, and fetch the band data objects.

```
rows_R_Band=Dataset_R_Band.RasterYSize
cols_R_Band=Dataset_R_Band.RasterXSize
Band_R=Dataset_R_Band.GetRasterBand(1)
NoDataVal_R=Band_R.GetNoDataValue()
Band_NIR=Dataset_NIR_Band.GetRasterBand(1)
NoDataVal_NIR=Band_NIR.GetNoDataValue()
```

Fetch the driver used for opening the red band and get its short name (or code). In this test case, the short name of driver is *GTiff*.

```
driver=Dataset_R_Band.GetDriver()
format=driver.ShortName
```

New raster file can be created, if the format driver supports it. There are two general techniques for creating files, using *CreateCopy()* and *Create()* method. When *CreateCopy()* method is called on the format driver, the source dataset should be passed which needs to be copied. When *Create()* method is called on the driver, one needs to explicitly write all the metadata and raster data with separate calls. All drivers that support creating new files support the *CreateCopy()* method, while vice-versa is not always true. To determine if a particular format supports *Create()* or *CreateCopy()*, one should check the *DCAP_CREATE* and *DCAP_CREATECOPY* metadata on the format driver object. Note that a number of drivers are read-only and would not support *Create()* or *CreateCopy()*.

```
metadata=driver.GetMetadata()
if metadata.has_key(gdal.DCAP_CREATE) \
  and metadata[gdal.DCAP_CREATE]=='YES':
    print 'Driver %s supports Create() method.' %format
if metadata.has_key(gdal.DCAP_CREATECOPY) \
  and metadata[gdal.DCAP_CREATECOPY]=='YES':
    print 'Driver %s supports CreateCopy() method.' %format
```

For situation in which one does not want to export an existing file to a new file, it is generally necessary to use the *Create()* method (though some interesting options are possible through use of virtual files or in-memory files). The *Create()* method takes option list much like *CreateCopy()*, but the image size, number of bands and band type must be provided explicitly. Once the dataset is successfully created, all appropriate metadata and raster data must be written to the file. If number of bands is not specified, then *Create()* method takes default value as 1; while the default data type (*eType* parameter) is *GDT_Byte*.

```
File_NDVI='C:/Results/NDVI.tif'
outDataSet=driver.Create(utf8_path=File_NDVI,xsize=cols_R_Band, \
            ysize=rows_R_Band,bands=1,eType=gdalconst.GDT_Float32)
```

The created dataset does not contain any geographical information. The generated NDVI (calculated using equation 4-3) should have the same geographical location and projection information as the input data, which can be set using *SetGeoTransform() and SetProjection()* methods, respectively. As the NDVI ranges from -1 to +1, set no data value as -999 using *SetNoDataValue()*. This can be any other value outside closed set of [-1, +1]. NDVI is calculated as:

$$NDVI = \frac{(NIR\ band - Red\ band)}{(NIR\ band + Red\ band)} \tag{4-3}$$

```
outDataSet.SetGeoTransform(Dataset_R_Band.GetGeoTransform())
outDataSet.SetProjection(Dataset_R_Band.GetProjection())
outBand=outDataSet.GetRasterBand(1)
outBand.SetNoDataValue(-999)
```

As the raster can be of large size, so the processing of the images should preferably be carried out one row at a time. For this, initialise an empty NumPy array of 32-bit floating point data tye, having same number of columns as input band but row length as 1.

```
ndvi=numpy.empty((1,cols_R_Band),numpy.float32)
```

Process the data row wise using *for* loop, and keep pixel value of -999 where NDVI calculation fails.

```
for row in range(rows_R_Band):
```

ReadAsArray() method can be used to read raster band data as NumPy arrays, in this case, reading one row at a time. The parameters *xoff* and *yoff* takes the offset values to start reading at (default arguments of 0), while *win_xsize* and *win_ysize* accepts number of pixels to read in X and Y direction (default arguments of 0). The *astype()* method is used to convert the data to a specific data type.

```
    Data_R_Band=Band_R.ReadAsArray(xoff=0,yoff=row,win_xsize=cols_R_Band, \
                        win_ysize=1).astype(numpy.float32)
    Data_NIR_Band=Band_NIR.ReadAsArray(xoff=0,yoff=row,win_xsize=cols_R_Band, \
                        win_ysize=1).astype(numpy.float32)
```

NDVI is calculated using *select()* method, which return an array drawn from elements in *choicelist*, depending upon the conditions.

```
    condlist=[Data_R_Band==NoDataVal_R,Data_NIR_Band==NoDataVal_NIR, \
                        (Data_R_Band+Data_NIR_Band)==0]
    choicelist=[-999,-999,-999]
    ndvi=numpy.select(condlist,choicelist, \
    ((Data_NIR_Band-Data_R_Band)/(Data_NIR_Band+Data_R_Band)))
```

Python implementation of writing a chunk of GDAL file from a NumPy array can be achieved by *WriteArray()* method. The parameter *array* takes the data as array which needs to be written, *xoff* and *yoff* takes the offset values to start writing (default arguments of 0).

```
    outBand.WriteArray(array=ndvi,xoff=0,yoff=row)
```

Make all open datasets out of scope by assigning to *None*.

```
Dataset_R_Band=Dataset_NIR_Band=outDataSet=None
```

4.4. REPROJECTION AND RESAMPLING

Take a step further to undertake a task of reprojecting and resampling multi-band satellite imagery (*SatImage.tif*). Start with importing the required modules/packages.

```
#!/usr/bin/env python
import osgeo.gdal as gdal
import osgeo.gdal_array as gdal_array
import osgeo.osr as osr
import matplotlib.pyplot as plt
import math,numpy
```

Fetch the GeoTIFF GDAL driver based on the short name, followed by registering the driver.

```
driver=gdal.GetDriverByName('GTiff')
driver.Register()
```

Specify the TIFF filename as input which needs to be reprojected and outfile file name after reprojection.

```
inFile='C:/Data/SatImage.tif'
outFile='C:/Results/ReprojectedImg.tif'
```

Open the input file in read-only mode, and read the whole dataset as NumPy array.

```
inDataset=gdal.Open(inFile,0)
inRasterData=gdal_array.DatasetReadAsArray(inDataset)
```

Also get hold of SRS, raster size, data type, transformation parameters, and calculate corner coordinates of input data.

```
in_srs=osr.SpatialReference()
in_srs.ImportFromWkt(inDataset.GetProjection())
in_DataType=inDataset.GetRasterBand(1).DataType
in_col=inDataset.RasterXSize
in_row=inDataset.RasterYSize
in_band=inDataset.RasterCount
in_geo=inDataset.GetGeoTransform()
in_pixelX,in_pixelY=in_geo[1],-in_geo[5]
in_ulx,in_uly=in_geo[0],in_geo[3]
in_urx,in_ury=in_ulx+in_pixelX*in_col,in_uly
in_lrx,in_lry=in_urx,in_uly-in_pixelY*in_row
in_llx,in_lly=in_ulx,in_lry
```

Create an SRS instance and assign *WGS 84 / UTM zone 16N* coordinate system using EPSG code.

```
out_srs=osr.SpatialReference()
out_srs.ImportFromEPSG(32616)
```

Initialise a *CoordinateTransformation* class instance with arguments as source and target SRS instances, followed by coordinate transformation using *TransformPoint()* method.

```
tx=osr.CoordinateTransformation(in_srs,out_srs)
(out_ulx,out_uly,out_ulz)=tx.TransformPoint(in_ulx,in_uly)
(out_urx,out_ury,out_urz)=tx.TransformPoint(in_urx,in_ury)
(out_lrx,out_lry,out_lrz)=tx.TransformPoint(in_lrx,in_lry)
(out_llx,out_lly,out_llz)=tx.TransformPoint(in_llx,in_lly)
```

There is a relation between pixel size and number of rows/columns, so either specify pixel size and calculate the row/column numbers, or vice-versa. Also calculate the spatial extent and transformation parameters of reprojected image.

```
minX=min(out_ulx,out_llx)
maxX=max(out_urx,out_lrx)
minY=min(out_lly,out_lry)
maxY=max(out_ury,out_uly)
opt=raw_input('''Enter 1 for specifying rows/columns of reprojected raster,
        2 for specifying pixel resolution of reprojected raster: ''')
if int(opt)==1:
  out_row=int(raw_input('Enter number of rows of reprojected raster: '))
  out_col=int(raw_input('Enter number of columns of reprojected raster: '))
elif int(opt)==2:
  out_pixelX=float(raw_input('Enter pixel width of reprojected raster: '))
  out_pixelY=float(raw_input('Enter pixel height of reprojected raster: '))
  out_col=int(math.floor((maxX-minX)/out_pixelX))
  out_row=int(math.floor((maxY-minY)/out_pixelY))
out_pixelX=(maxX-minX)/out_col
out_pixelY=(maxY-minY)/out_row
out_geo=(minX,out_pixelX,0,maxY,0,-out_pixelY)
```

GDAL supports an option of holding raster in a temporary in-memory format. This is primarily useful for temporary datasets in scripts or internal to applications. In-memory dataset supports auxilary information including metadata, coordinate systems, georeferencing, GCPs, colour interpretation, no data value, colour tables and all pixel data types. Create an in-memory raster followed by setting projection, geotransformation and no data information.

```
mem_drv=gdal.GetDriverByName('MEM')
outDataset=mem_drv.Create('',out_col,out_row,in_band,in_DataType)
outDataset.SetGeoTransform(out_geo)
outDataset.SetProjection(out_srs.ExportToWkt())
for iBand in range(inDataset.RasterCount):
  in_band=inDataset.GetRasterBand(iBand+1)
  NoDataVal=in_band.GetNoDataValue()
  outDataset.GetRasterBand(iBand+1).SetNoDataValue(NoDataVal)
```

A convenience way to reproject an image is by using *ReprojectImage()* method. In particular, this method takes care of establishing the transformation function to implement the reprojection, but no metadata, projection information, or colour tables are transferred to the output file. One can choose a resampling algorithm from table 4-8, but the method uses *GRA_NearestNeighbour* as default argument.

Table 4-8: Resampling algorithms

	Resampling algorithm	Description
Numeric	String	
0	*GRA_NearestNeighbour*	Nearest neighbour (select on one input pixel)
1	*GRA_Bilinear*	Bilinear (2x2 kernel)
2	*GRA_Cubic*	Cubic Convolution Approximation (4x4 kernel)
3	*GRA_CubicSpline*	Cubic B-Spline Approximation (4x4 kernel)
4	*GRA_Lanczos*	Lanczos windowed sinc interpolation (6x6 kernel)

```
gdal.ReprojectImage(inDataset,outDataset,in_srs.ExportToWkt(), \
            out_srs.ExportToWkt(),gdal.GRA_NearestNeighbour)
```

In-memory raster can be saved to any format supported by GDAL by making a copy of the dataset using *CreateCopy()*. It returns a writeable dataset, and it must be closed properly to complete writing and flushing the dataset to disk. In the case of Python, this occurs automatically when output dataset goes out of scope. Band number, size, type, projection, geotransform and so forth are all to be copied from the provided template dataset. The passing of *False* (or 0) argument to *strict* parameter (default argument *True*) indicates that the *CreateCopy()* call should proceed without a fatal error even if the destination dataset cannot be created to exactly match the input dataset. This might be because the output format does not support the pixel data type of the input dataset, or because the destination cannot support writing georeferencing for instance.

```
dst_ds=driver.CreateCopy(utf8_path=outFile,src=outDataset,strict=0)
```

Plot original and reprojected image, and then close the datasets.

```
inRasterData=numpy.rollaxis(inRasterData,0,3)
fig1=plt.figure(figsize=(9,7),facecolor='white')
fig1.canvas.set_window_title("Original image")
ax1=fig1.add_subplot(1,1,1)
cax=ax1.imshow(X=inRasterData)
ax1.axis("off")
outRasterData=gdal_array.DatasetReadAsArray(outDataset)
outRasterData=numpy.rollaxis(outRasterData,0,3)
fig2=plt.figure(figsize=(9,7),facecolor='white')
fig2.canvas.set_window_title("Reprojected image")
ax2=fig2.add_subplot(1,1,1)
cax=ax2.imshow(X=outRasterData)
ax2.axis("off")
inDataset=outDataset=dst_ds=fig1=fig2=None
```

Open the reprojected image in read only mode and use *BuildOverviews()* method to build raster overviews. The resampling parameter is one of *NEAREST*, *GAUSS*, *CUBIC*, *AVERAGE*, *MODE*, *AVERAGE_MAGPHASE* or *NONE*, for controlling the applied downsampling method, with *NEAREST* as the default argument. The *overviewlist* parameter takes a list of overview decimation factors to build. Some format drivers do not support overviews at all. Many format drivers store overviews in a secondary file with the extension *.ovr*, which is actually in TIFF format. Most drivers also support an alternate overview format using Erdas Imagine format *.rrd*.

```
Dataset=gdal.Open(outFile,gdal.GA_ReadOnly)
Dataset.BuildOverviews(resampling="NEAREST",overviewlist=[2,4,8,16,32])
```

Close the dataset after creating overviews.

```
Dataset=None
```

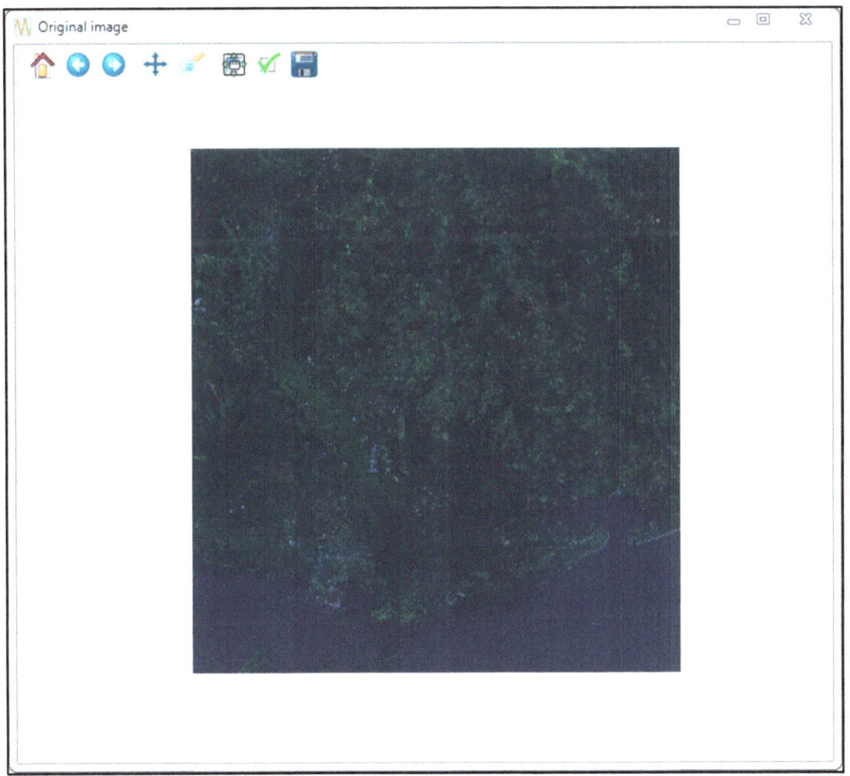

Figure 4-3: Satellite imagery (SatImage.tif) plot

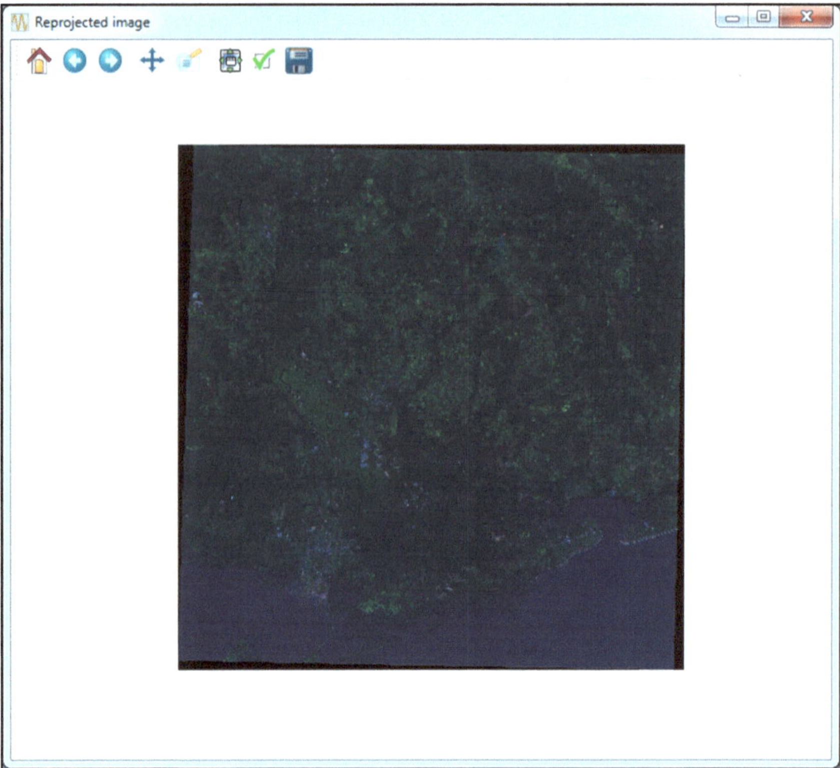

Figure 4-4: Reprojected image

4.5. VECTOR INFORMATION

Till now, handling of raster data has been discussed. Now move ahead for learning technique to handle vector data. The test data is a shapefile (*LKA_adm0.shp*) and the objective is to read (and extract useful information) and plot. Import OGR module for processing vector data information; and other required modules/classes.

```
#!/usr/bin/env python
try:
  import osgeo.ogr as ogr
except ImportError:
  import ogr
from mpl_toolkits.basemap import Basemap
import matplotlib.pyplot as plt
import sys,os
```

An OGR data source represents a set of OGR layer objects (discussed later). Initialise a variable with path of the input shapefile.

```
File='C:/Data/LKA_adm0.shp'
```

Open() method attempt to open the file with an OGR driver; *update* parameter as 0 or *False* (default argument) access the file in read-only mode. The method returns *None* on error or if the passed file format is not supported by driver, otherwise a handle to an OGR data source is returned.

```
DataSource=ogr.Open(File,update=0)
if DataSource is None:
  print 'Unable to open file'
  sys.exit(1)
```

We are already aware of the name of data source, but for the sake of learning new functions, print the name of the data source using *GetName()*.

```
print("Data source: %s" %(DataSource.GetName()))
```

Get hold of the driver that has been used to open the data source.

```
driver= DataSource.GetDriver()
```

Fetch name of the driver, which is relatively short (10-40 characters), and should reflect the underlying file format, for instance, *ESRI Shapefile*.

```
print("Driver: %s" %(driver.GetName()))
```

An OGR layer represents a layer of features within a data source. All features in an OGR layer share a common schema, and are of the same OGR feature definition. The OGR layer can be thought of as a gateway for reading and writing features from an underlying data source.

Now extract information stored in all layers of a given vector data. This can be accomplished by a loop (*for* loop) which runs through all OGR layers, extracting information of each layer one at a time. *GetLayerCount()* gets the number of layers in a data source.

```
for iLayer in range(Dataset.GetLayerCount()):
```

GetLayer() fetch a layer by index.

```
  layer= DataSource.GetLayer(iLayer)
```

GetLayerDefn() fetches the schema information for the layer. It returns OGR feature definition which is owned by the OGR layer, and it encapsulates the attribute schema of the features of the layer.

```
  layer_defn=layer.GetLayerDefn()
  print("")
```

GetName() returns the layer name.

print("Layer name: %s" %layer_defn.GetName())

GetGeomType() return the layer geometry type. Table 4-9 enlists Well Known Binary (WKB) format of geometry types.

Table 4-9: OGR layer geometry types

WKB geometry type	Description
wkbUnknown	unknown type, non-standard
wkbPoint	0-dimensional geometric object, standard WKB
wkbLineString	1-dimensional geometric object with linear interpolation between Points, standard WKB
wkbPolygon	planar 2-dimensional geometric object defined by 1 exterior boundary and 0 or more interior boundaries, standard WKB
wkbMultiPoint	GeometryCollection of Points, standard WKB
wkbMultiLineString	GeometryCollection of LineStrings, standard WKB
wkbMultiPolygon	GeometryCollection of Polygons, standard WKB
wkbGeometryCollection	geometric object that is a collection of 1 or more geometric objects, standard WKB
wkbNone	non-standard, for pure attribute records
wkbLinearRing	non-standard, just for createGeometry()
wkbPoint25D	2.5D extension as per 99-402
wkbLineString25D	2.5D extension as per 99-402
wkbPolygon25D	2.5D extension as per 99-402
wkbMultiPoint25D	2.5D extension as per 99-402
wkbMultiLineString25D	2.5D extension as per 99-402
wkbMultiPolygon25D	2.5D extension as per 99-402
wkbGeometryCollection25D	2.5D extension as per 99-402

where the reference 99-402 is an OGC document number. An HTML version of this documnet is available at *http://home.gdal.org/projects/opengis/twohalfdsf.html.*

GeometryTypeToName() fetches a human readable name corresponding to an OGR WKB geometry type value.

print("Geometry: %s" %ogr.GeometryTypeToName(layer_defn.GetGeomType()))

GetFeatureCount() method fetch the number of features in a layer. For dynamic databases, the count may not be exact. If *force* parameter is 0 (or *False*) and if it is expensive to establish the feature count, a value of -1 may be returned indicating that the count is not known. If *force* is 1 (default value), some implementation will actually scan the entire layer once to count objects.

print("Feature count: %d" %layer.GetFeatureCount(force=1))

GetExtent() fetch the Minimum Bounding Rectangle (MBR) of the data in the layer. If *force* parameter is *False*, and if it is expensive to establish the extent, then *None* will be returned indicating that the extent is not known. If *force* is *True*, then some implementations will actually scan the entire layer once to compute the MBR of all the features in the layer. Layers without any geometry may return *None*, just indicating that no meaningful extents could be collected.

```
oExt=layer.GetExtent(True)
if oExt is not None:
    print("Extent: (%f, %f) - (%f, %f)" %(oExt[0],oExt[1],oExt[2],oExt[3]))
```

GetSpatialRef() fetches the spatial reference system for this layer. On success, the method returns spatial reference, or *None* if there is not one.

```
if layer.GetSpatialRef() is None:
    srs_WKT="(unknown)"
else:
    srs_WKT=layer.GetSpatialRef().ExportToPrettyWkt()
print("Layer SRS WKT:\n%s" %srs_WKT)
```

The feature ID (FID) of a feature is intended to be a unique identifier for the feature within the layer. The feature ID is modelled in OGR as long integer; however, this is not sufficiently expressive to model the natural feature ID in some formats. For instance, the GML feature ID is a string. *GetFIDColumn()* method returns the name of the underlying database column being used as the FID column, or empty string ("") if not supported.

```
if len(layer.GetFIDColumn())>0:
    print("FID Column: %s" %layer.GetFIDColumn())
```

GetGeometryColumn() method returns the name of the underlying database column being used as geometry column, or empty string ("") if not supported.

```
if len(layer.GetGeometryColumn())>0:
    print("Geometry Column: %s" %layer.GetGeometryColumn())
```

Extract information stored in all fields of a feature in a given vector data layer using a loop (*for* loop). *GetFieldCount()* fetches the number of fields in the passed feature definition.

```
for iAttr in range(layer_defn.GetFieldCount()):
```

Fetch the field definition using *GetFieldDefn()* method for each feature.

```
    Field=layer_defn.GetFieldDefn(iAttr)
```

Name of the field can be obtained by *GetNameRef()* method. *GetType()* returns the feature field type; the list of field types are given in table 4-10.

Table 4-10: Types of OGR feature fields

OGR feature field types		Description
Numeric	String	
0	OFTInteger	Simple 32-bit integer
1	OFTIntegerList	List of 32-bit integers
2	OFTReal	Double Precision floating point
3	OFTRealList	List of doubles
4	OFTString	String of ASCII chars
5	OFTStringList	Array of strings
6	OFTWideString	deprecated
7	OFTWideStringList	deprecated
8	OFTBinary	Raw Binary data
9	OFTDate	Date
10	OFTTime	Time
11	OFTDateTime	Date and Time

Human readable name for a field type can be obtained by *GetFieldTypeName()* method. *GetWidth()* get the formatting width for this field; returns 0 when there is no specified width. *GetPrecision()* gets the formatting precision for this field and it should normally be 0 for fields of types other than *OFTReal*.

```
print("%s: %s (%d.%d)" % ( \
        Field.GetNameRef(), \
        Field.GetFieldTypeName(Field.GetType()), \
        Field.GetWidth(), \
        Field.GetPrecision()))
```

GetNextFeature() method implements sequential access to the next feature of a layer. Pass on this feature to a user-defined function *FeatureInfo()*, which will extract the contents of the feature under consideration. Run a loop (*while* loop) to extract information from all features.

```
Feature=layer.GetNextFeature()
while Feature is not None:
  FeatureInfo(Feature)
  Feature=layer.GetNextFeature()
```

Set up a basemap with transverse mercator map projection.

```
m = Basemap(projection='tmerc',lon_0=80.7,lat_0=7.9, \
      k_0=0.9996,rsphere=(6378137.00,6356752.314245179), \
      width=320000,height=480000,resolution='i')
```

Get the complete path of the shapefile without its extension (*.shp*).

```
(filepath,filename)=os.path.split(File)
```

```
(shortname,extension)=os.path.splitext(filename)
shapefile=filepath+'/'+shortname
```

readshapefile() method is used to read shapefile with an option to draw map boundary. The method assumes that the shapes are 2D and it only works for point, multipoint, polyline and polygon shapes. Please note that the vertices must be in geographic (latitude, longitude) coordinates. The *shapefile* parameter takes path to shapefile components as argument. Example: *shapefile='/home/jeff/esri/world_borders'* assumes that *world_borders.shp*, *world_borders.shx* and *world_borders.dbf* resides in */home/jeff/esri*. *name* parameter takes name for Basemap attribute to hold the shapefile vertices or points in map projection coordinates. The following optional keyword arguments are only relevant for polyline and polygon shape types; for point and multipoint shapes they are ignored. *drawbounds* parameter draw boundaries of shapes (default *True*), *linewidth* defines shape boundary line width (default 0.5), *color* takes shape boundary line colour (default *black*) argument, while *antialiased* parameter takes antialiasing switch for shape boundaries (default *True*).

```
s=m.readshapefile(shapefile,name='admin',drawbounds=True,linewidth=0.5, \
        color='k',antialiased=1)
```

Assign a title to the plot; the plot is shown in figure 4-5.

```
plt.title("Sri Lanka",fontsize=15)
```

The data source should be closed by assigning as *None*.

```
DataSource=m=None
```

The definition of *FeatureInfo()* function which is used to extract information of a feature is discussed below.

```
def FeatureInfo(Feature):
```

Fetch feature definition using *GetDefnRef()*.

```
  feature_Defn=Feature.GetDefnRef()
```

Get the name of the OGR feature definition using *GetName()*; and fetch feature identifier using *GetFID()*.

```
  print("OGRFeature(%s): %ld" %(feature_Defn.GetName(),Feature.GetFID()))
```

Manoeuvre through the fields of a feature and get field names and field types.

```
  for iField in range(feature_Defn.GetFieldCount()):
    field_Defn=feature_Defn.GetFieldDefn(iField)
```

```
line=" %s (%s) : " %(field_Defn.GetNameRef(), \
        ogr.GetFieldTypeName(field_Defn.GetType()))
```

IsFieldSet() method test if a field has ever been assigned a value or not. If yes, *GetFieldAsString()* is used to fetch field value as a string.

```
if Feature.IsFieldSet(iField):
  line=line+"%s" %(Feature.GetFieldAsString(iField))
else:
  line=line+"(null)"
print(line)
```

GetStyleString() fetches style string (i.e. colours, line width, symbols, etc.) for a feature. A reference to a representation in string format is returned, if present, or *None* is returned. To know more about feature style, visit the website link *http://www.gdal.org/ogr/ogr_feature_style.html.*

```
if Feature.GetStyleString() is not None:
  print("Style: %s" %Feature.GetStyleString())
```

Obtain a handle to feature geometry using *GetGeometryRef()*, and pass it to a user defined method called *GeometryInfo()* to extract information of given feature geometry.

```
Geometry=Feature.GetGeometryRef()
if Geometry is not None:
  GeometryInfo(Geometry)
print('')
return
```

```
def GeometryInfo(Geometry):
```

Fetch WKT name for geometry type using *GetGeometryName()*.

```
line=("%s%s : " %(" ",Geometry.GetGeometryName()))
```

GetGeometryType() aquires the geometry type in WKB format.

```
eType=Geometry.GetGeometryType()
```

Get the number of rings information (if present), and/or the number of point count of geometry using the following patch of code.

```
if eType==ogr.wkbLineString or eType==ogr.wkbLineString25D:
  line=line+("%d points" %Geometry.GetPointCount())
  print(line)
elif eType==ogr.wkbPolygon or eType==ogr.wkbPolygon25D:
```

```
    nRings=Geometry.GetGeometryCount()
    if nRings==0:
        line=line+"empty"
    else:
        poRing=Geometry.GetGeometryRef(0)
        line=line+("%d points" %poRing.GetPointCount())
        if nRings>1:
            line=line+(", %d inner rings (" %(nRings-1))
            for ir in range(0,nRings-1):
                if ir>0:
                    line=line+", "
                poRing=Geometry.GetGeometryRef(ir+1)
                line=line+("%d points" %poRing.GetPointCount())
            line=line+")"
    print(line)
elif eType==ogr.wkbMultiPoint or \
    eType==ogr.wkbMultiPoint25D or \
    eType==ogr.wkbMultiLineString or \
    eType==ogr.wkbMultiLineString25D or \
    eType==ogr.wkbMultiPolygon or \
    eType==ogr.wkbMultiPolygon25D or \
    eType==ogr.wkbGeometryCollection or \
    eType==ogr.wkbGeometryCollection25D:
        line=line+"%d geometries:" %Geometry.GetGeometryCount()
        print(line)
        for ig in range(Geometry.GetGeometryCount()):
            subgeom=Geometry.GetGeometryRef(ig)
            from sys import version_info
            if version_info>=(3,0,0):
                exec('print("", end=" ")')
            else:
                exec('print "", ')
            GeometryInfo(subgeom)
else:
    print(line)
return
```

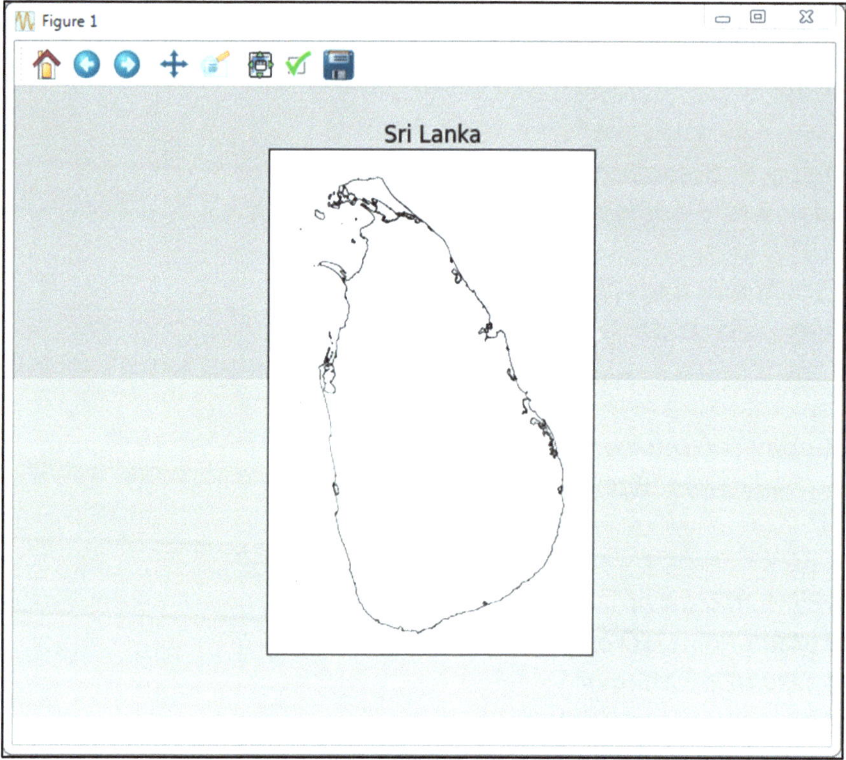

Figure 4-5: Shapefile (LKA_adm0.shp) plot

4.6. CONTOUR GENERATION

The script discussed in this section generates contour shapefile using Digital Elevation Model (DEM) data. The test data is *srtm_45_07.tif*, which is basically an SRTM data. More information on this data is given in chapter 6. Start the script by importing required modules, followed by opening the given SRTM file.

```
#!/usr/bin/env python
import osgeo.gdal as gdal
import osgeo.ogr as ogr
import osgeo.osr as osr
import matplotlib.pyplot as plt
import sys,os,time
infile=r'C:\Data\srtm_45_07.tif'
dataset=gdal.Open(infile)
if dataset is None:
   print 'Unable to open file'
   sys.exit(1)
```

Obtain the first (and only) band of SRTM data, followed by initializing a spatial reference object with same SRS as input.

```
band=dataset.GetRasterBand(1)
```

```
spatial_ref=osr.SpatialReference()
spatial_ref.ImportFromWkt(dataset.GetProjection())
```

Obtain no data value of the band and transformation parameters of the input dataset.

```
NoDataVal=band.GetNoDataValue()
geotrans=dataset.GetGeoTransform()
```

Ask the user for contour shapefile name (including path) and the field name in which contour elevation information will be stored.

```
outfile=raw_input('Enter output contour shapefile name (including path): ')
field_name=raw_input('Enter field name for elevation data in contour shapefile: ')
```

Create a new OGR data source based on passed *ESRI Shapefile* driver using *CreateDataSource()* method.

```
ogr_ds=ogr.GetDriverByName('ESRI Shapefile').CreateDataSource(outfile)
```

CreateLayer() method attempts to create a new layer on the data source with the indicated name, coordinate system, and geometry type(*wkbUnknown* as default).

```
(filepath,filename)=os.path.split(outfile)
(shortname,extension)=os.path.splitext(filename)
ogr_lyr=ogr_ds.CreateLayer(str(shortname),spatial_ref)
```

Create an ID field of interger data type.

```
field_defn=ogr.FieldDefn('ID',ogr.OFTInteger)
ogr_lyr.CreateField(field_defn)
```

Create a floating data type field for storing contour elevation.

```
field_defn=ogr.FieldDefn(str(field_name),ogr.OFTReal)
ogr_lyr.CreateField(field_defn)
```

Now contours can be generated via two different approaches. In the first approach, contours of specific elevations will be generated, while another approach will generate contours for given contour interval. The *ContourGenerate()* method is used create contour vector data from raster DEM. The algorithm generates contour vector data for the input raster band on the requested set of contour levels. Then the contours are written to the passed in OGR vector layer. Also, a no data value may be specified to identify pixels that should not be considered in contour line generation. Following is a discussion of parameters taken by method:
srcBand: The band to read raster data from; the whole band will be processed.

contourInterval: The elevation interval between contours generated.
contourBase: The base relative to which contour intervals are applied. This is normally zero, but could be different. To generate 10 m contours at 5, 15, 25, and so on, the *ContourBase* should be 5.
fixedLevelCount: The list of fixed contour levels at which contours should be generated. If this is greater than zero, then fixed levels will be used, and *contourInterval* and *contourBase* are ignored.
useNoData: If *True*, the no data value will be used.
noDataValue: The value to use as a no data. This is a pixel value which should be ignored in generating contours, as if the value of the pixel were not known.
dstLayer: The layer to which generated contours will be written.
idField: If not -1, this will be used as a field index to indicate where a unique ID should be written for each feature (contour).
elevField: If not -1, this will be used as a field index to indicate where the elevation value of the contour should be written.

```
condition=raw_input('Enter 1 for specific contour generation; else 2 for \
            generating contours based on contour interval: ')
if int(condition) is 1:
  FixedLevels=raw_input('Enter space separated elevation data: ')
  t1=time.clock()
  FixedLevels=[float(x) for x in FixedLevels.split()]
  gdal.ContourGenerate(srcBand=band,contourInterval=0,contourBase=0, \
  fixedLevelCount=FixedLevels,useNoData=0,noDataValue=float(NoDataVal), \
  dstLayer=ogr_lyr,idField=0,elevField=1)
  t2=time.clock()
elif int(condition) is 2:
  contour_intvl=raw_input('Enter contour interval: ')
  BaseContourValue=raw_input('Enter elevation of base contour: ')
  t1=time.clock()
  gdal.ContourGenerate(band,float(contour_intvl),float(BaseContourValue), \
             [],0,float(NoDataVal),ogr_lyr,0,1)
  t2=time.clock()
print 'Contours were generated in %f seconds'%(t2-t1)
```

Calculate band statistics and get hold of band data.

```
min=band.GetMinimum()
max=band.GetMaximum()
if min is None or max is None:
  MinMax=band.ComputeRasterMinMax(False)
  min=MinMax[0]
  max=MinMax[1]
BandData=band.ReadAsArray()
```

Display SRTM data using Matplotlib library (figure 4-6).

```
fig1=plt.figure(figsize=(9,7),facecolor='white')
fig1.canvas.set_window_title("SRTM data plot")
ax1=fig1.add_subplot(1,1,1)
```

Prepare an opaque (*alpha* parameter takes value between 0 (transparent) and 1 (opaque)) plot of the SRTM data between *vmin*, *vmax* values (used to scale an image to 0-1), using *terrain* colourmap (refer figure 3-1 for colourmaps).

```
cax=ax1.imshow(X=BandData,vmin=min,vmax=max,alpha=1,cmap=plt.cm.terrain)
```

Add a vertical oriented (can also be horizontal) colourbar to the plot, with ticks plotted at minimum and maximum data values.

```
cbar=fig1.colorbar(cax,ticks=[min,max],orientation='vertical')
cbar.ax.set_yticklabels([str(min),str(max)])
```

Label the axes and plot axes ticks.

```
ax1.set_xlabel('Longitude (degree)')
ax1.set_ylabel('Latitude (degree)')
ax1.set_title("SRTM data plot")
ax1.set_xticks((0,1200,2400,3600,4800,dataset.RasterXSize))
x=int(geotrans[0])
ax1.set_xticklabels((x,x+1,x+2,x+3,x+4,x+5))
ax1.set_yticks((0,1200,2400,3600,4800,dataset.RasterYSize))
y=int(geotrans[3])
ax1.set_yticklabels((y,y-1,y-2,y-3,y-4,y-5))
ax1.grid(True)
```

Initialise a figure (with a subplot) and assign a title, for plotting histogram (figure 4-7).

```
pix=dataset.RasterXSize*dataset.RasterYSize
fig2=plt.figure(figsize=(9,7),facecolor='white')
fig2.canvas.set_window_title("Histogram of SRTM data")
ax2=fig2.add_subplot(111)
```

Convert 2-D array data to 1-D array.

```
BandData=BandData.reshape(pix,)
```

Compute and draw the histogram of band data using *hist()* method, with the following parameters:
x- Band data used to plotting histogram. Multiple data can be provided as a list of datasets of potentially different length ([*x0*, *x1*, ...]), or as a 2-D array, in which each column is a dataset.

bins- Either an integer number of bins or a sequence giving the bins. If *bins* is an integer, *bins* + 1 bin edges will be returned. Unequally spaced bins are supported if *bins* is a sequence.
range- The lower and upper range of the bins. If not provided, *range* is (*x.min(), x.max()*). Range has no effect if *bins* is a sequence.
alpha- Accepts float value (0.0 transparent through 1.0 opaque).
histtype- The type of histogram to draw.

- *bar* is a traditional bar-type histogram. If multiple data are given the bars are aranged side by side.
- *barstacked* is a bar-type histogram where multiple data are stacked on top of each other.
- *step* generates a lineplot that is by default unfilled.
- *stepfilled* generates a lineplot that is by default filled.

align- Controls how the histogram is plotted.

- *left*: bars are centred on the left bin edges.
- *mid*: bars are centred between the bin edges.
- *right*: bars are centred on the right bin edges.

orientation- The possible arguments are *horizontal* and *vertical.*

```
ax2.hist(x=BandData,bins=100,range=(min,max),facecolor='green',alpha=1.0, \
    histtype='bar',align='mid',orientation='vertical')
```

Set axis labels and plot title.

```
ax2.set_xlabel('Elevation (meter)')
ax2.set_ylabel('Frequency')
ax2.set_title('Histogram of SRTM data')
ax2.set_xlim(min,max)
ax2.grid(True)
```

Destroy objects by assigning them as *None*.

```
dataset=fig=None
ogr_ds.Destroy()
```

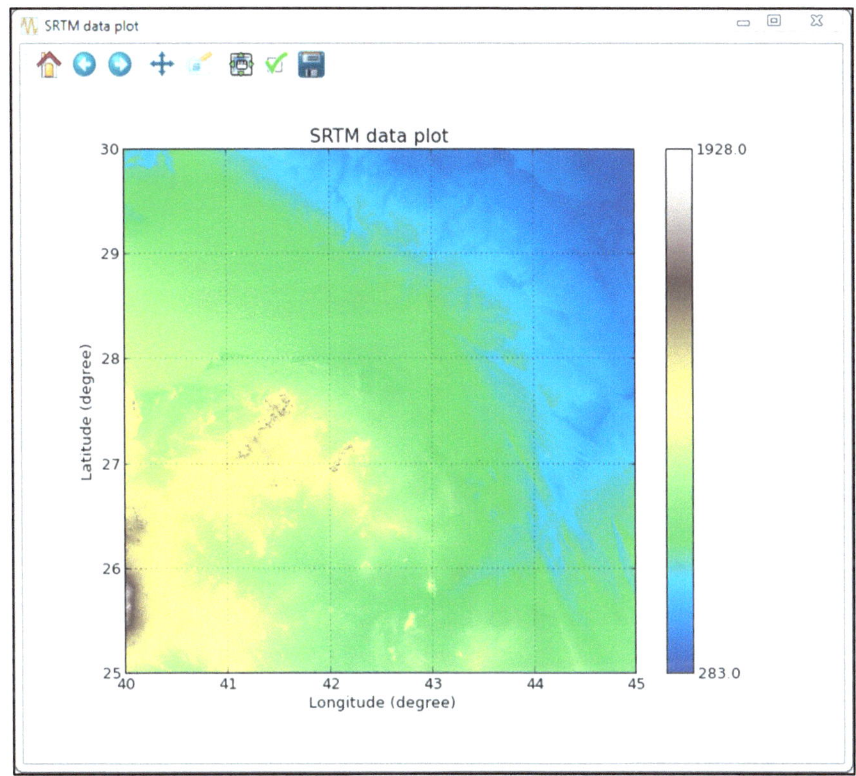

Figure 4-6: SRTM data (srtm_45_07.tif) plot

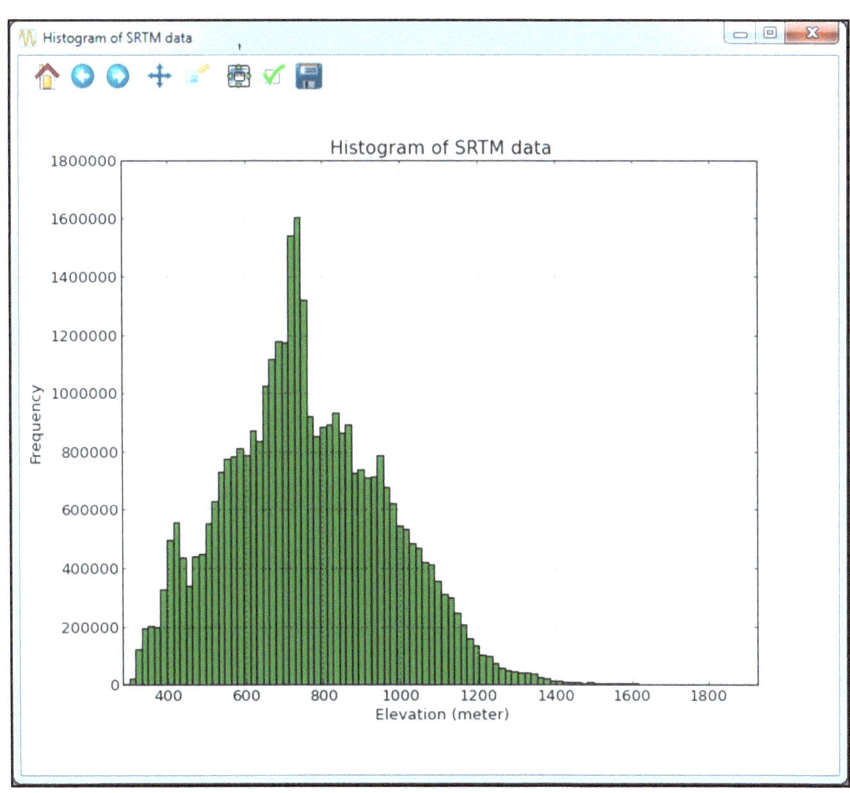

Figure 4-7: Histogram of SRTM data (srtm_45_07.tif)

4.7. RASTER CLIPPING

The focus of next script is on raster data (*SatImage.tif*) clipping operation using vector data (*county.shp*). In this process, also generate a mask band which has spatial extent as vector data. Start with importing required modules and register all GDAL drivers.

```
#!/usr/bin/env python
import osgeo.gdal as gdal
import osgeo.gdal_array as gdal_array
import osgeo.gdalconst as gdalconst
import osgeo.ogr as ogr
import osgeo.osr as osr
import math,numpy
import matplotlib.pyplot as plt
gdal.AllRegister()
```

Initialise variables with paths of raster data which needs to be clipped, the shapefile which is used for clipping, the output raster mask and clipped raster.

```
inRaster='C:/Data/SatImage.tif'
inVector='C:/Data/county.shp'
outMask='C:/Results/outMask.tif'
outRaster='C:/Results/outRaster.tif'
```

Open the raster file using *GTiff* driver and extract the required information.

```
driver=gdal.GetDriverByName('GTiff')
inDataset=gdal.Open(inRaster,gdalconst.GA_ReadOnly)
inRasterData=gdal_array.DatasetReadAsArray(inDataset)
in_r_srs=osr.SpatialReference()
in_r_srs.ImportFromWkt(inDataset.GetProjection())
in_DataType=inDataset.GetRasterBand(1).DataType
in_r_geo=inDataset.GetGeoTransform()
in_r_col=inDataset.RasterXSize
in_r_row=inDataset.RasterYSize
in_r_band=inDataset.RasterCount
in_r_ulx,in_r_uly=in_r_geo[0],in_r_geo[3]
in_r_pixelX,in_r_pixelY=in_r_geo[1],-in_r_geo[5]
```

Open the shapefile, find the spatial extent, and calculate the offset between raster data and vector in XY direction in terms of number of pixels.

```
inShapefile=ogr.GetDriverByName('ESRI Shapefile').Open(inVector,0)
in_v_minX,in_v_maxX,in_v_minY,in_v_maxY=inShapefile.GetLayer(0).GetExtent()
offsetX=math.floor((in_v_minX-in_r_ulx)/in_r_pixelX)
```

```
out_r_ulx=in_r_ulx+in_r_pixelX*offsetX
offsetY=math.floor((in_r_uly-in_v_maxY)/in_r_pixelY)
out_r_uly=in_r_uly-in_r_pixelY*offsetY
out_r_geo=(out_r_ulx,in_r_pixelX,0,out_r_uly,0,-in_r_pixelY)
out_r_col=int(math.ceil((in_v_maxX-out_r_ulx)/in_r_pixelX))
out_r_row=int(math.ceil((out_r_uly-in_v_minY)/in_r_pixelY))
```

Initialise an in-memory raster and set relevant information in it. This raster will be used as a mask band.

```
mask=gdal.GetDriverByName('MEM').Create('',out_r_col,out_r_row,1, \
                        gdalconst.GDT_Byte)
mask.SetGeoTransform(out_r_geo)
mask.SetProjection(in_r_srs.ExportToWkt())
mask_band=mask.GetRasterBand(1)
mask_band.Fill(0)
```

RasterizeLayer() method rasterize all the geometric objects from a layer into a raster dataset. The output raster may be of any GDAL supported data type, though currently the burning is done either as *GDT_Byte* or *GDT_Float32*. This may be improved in the future. The parameter *dataset* takes output dataset (must be opened in update mode); *bands* take the list of bands to be updated; *layer* takes the layer to burn in; *burn_values* accepts a list of values to burn into the raster. Apart from above mentioned parameters, there are *options* (few options are discussed here) for controlling rasterization:

ATTRIBUTE- Identifies an attribute field on the features to be used for a burn in value. The value will be burned into all output bands. If specified, layer burn values will not be used.

ALL_TOUCHED- May be set to *TRUE* to set all pixels touched by the line or polygons, not just those whose centre is within the polygon or that are selected by brezenhams line algorithm. Default value is *FALSE*.

```
gdal.RasterizeLayer(dataset=mask,bands=[1],layer=inShapefile.GetLayer(0), \
          burn_values=[255],options=["ALL_TOUCHED=TRUE"])
```

Read the mask band data as NumPy array, and save this array in file of *GTiff* format with the prototype of mask. In background, *SaveArray()* basically uses *CreateCopy()* method which copies band, size, type, projection, geotransform and so forth, from the provided prototype dataset.

```
mask_data=mask_band.ReadAsArray(0,0,out_r_col,out_r_row)
gdal_array.SaveArray(mask_data,outMask,format="GTiff",prototype=mask)
```

Trim the input dataset having same spatial extent as that of mask.

```
inRasterData_trim=inRasterData[:,offsetY:(offsetY+out_r_row), \
                  offsetX:(offsetX+out_r_col)]
```

Create an in-memory dataset for clipped raster and assign appropriate information.

```
outDataset=gdal.GetDriverByName('MEM').Create('',out_r_col,out_r_row, \
                                    in_r_band,in_DataType)
outDataset.SetGeoTransform(out_r_geo)
outDataset.SetProjection(in_r_srs.ExportToWkt())
for iBand in range(outDataset.RasterCount):
   in_band=inDataset.GetRasterBand(iBand+1)
   NoDataVal=in_band.GetNoDataValue()
   outDataset.GetRasterBand(iBand+1).SetNoDataValue(NoDataVal)
outRasterData=gdal_array.DatasetReadAsArray(outDataset)
```

There are two approaches for raster clipping; one is clipping to minimum bounding rectangle of vector layer extent, while other is clipping to actual vector layer shape.

```
opt=raw_input('''Enter 1 for raster clipping to feature extent,
          2 for raster clipping to feature shape: ''')
if int(opt)==1:
   outRasterData=numpy.copy(inRasterData_trim)
elif int(opt)==2:
   condlist=[mask_data==0]
   choicelist=[int(NoDataVal)]
   outRasterData=numpy.copy(inRasterData_trim)
   outRasterData=numpy.select(condlist,choicelist,inRasterData_trim)
gdal_array.SaveArray(outRasterData,outRaster,format="GTiff",prototype=outDataset)
```

Plot original and clipped rasters (refer figure 4-8 to 4-10).

```
inRasterData=numpy.rollaxis(inRasterData,0,3)
fig1=plt.figure(figsize=(9,7),facecolor='white')
fig1.canvas.set_window_title("Original data")
ax1=fig1.add_subplot(1,1,1)
cax=ax1.imshow(X=inRasterData)
ax1.axis("off")
outRasterData=numpy.rollaxis(outRasterData,0,3)
fig2=plt.figure(figsize=(9,7),facecolor='white')
fig2.canvas.set_window_title("Clipped data")
ax2=fig2.add_subplot(1,1,1)
cax=ax2.imshow(X=outRasterData)
ax2.axis("off")
```

After clipping, all dataset handles should be made out of scope.

```
inDataset=outDataset=mask=None
```

inShapefile.Destroy()

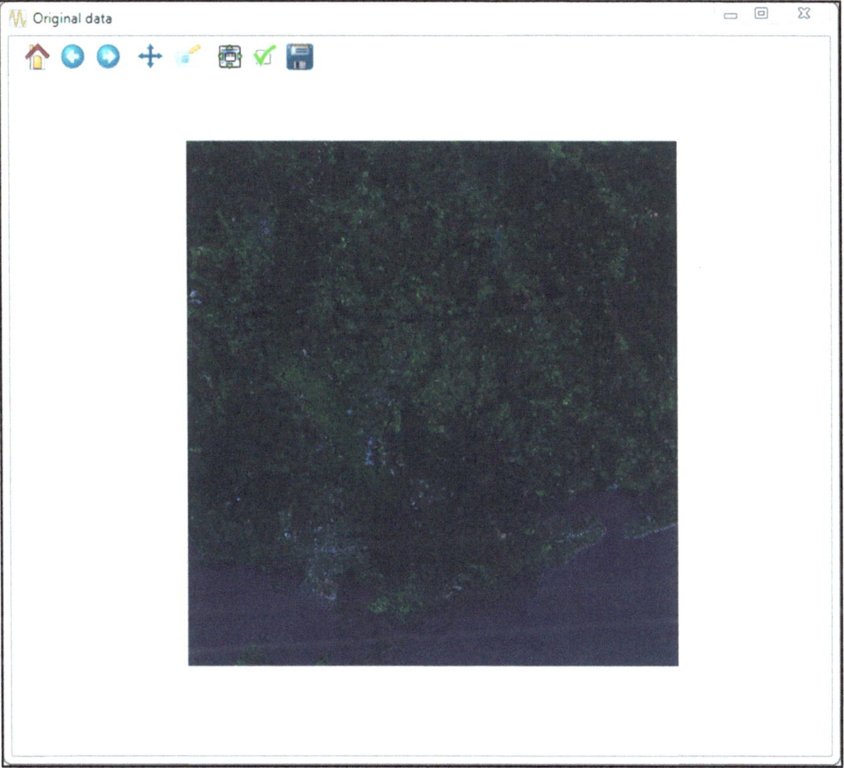

Figure 4-8: Satellite imagery (SatImage.tif) plot

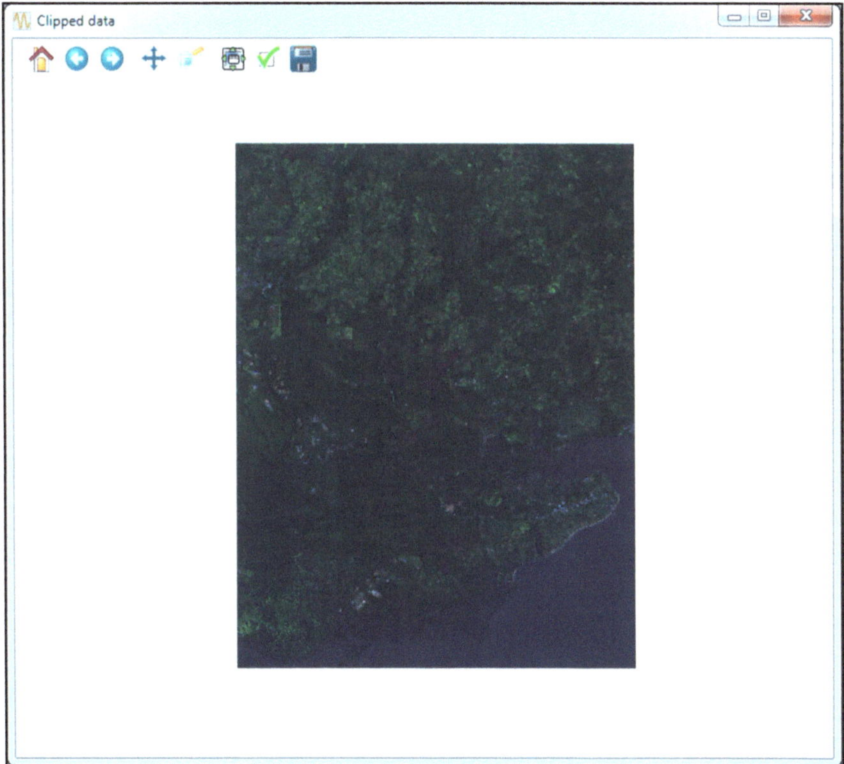

Figure 4-9: Raster clipped to extent

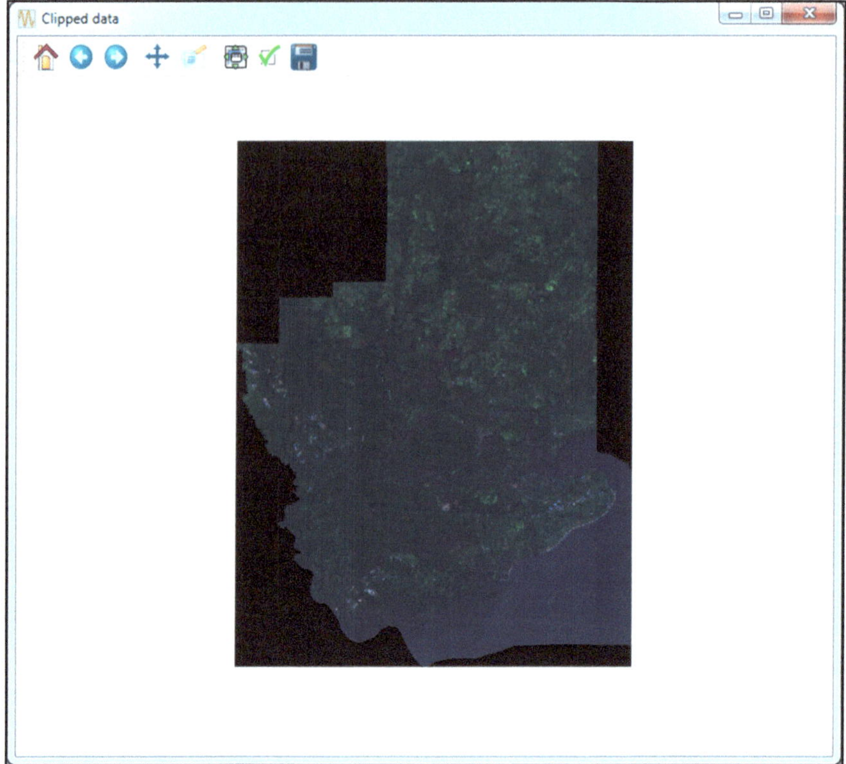

Figure 4-10: Raster clipped to vector shape

4.8. GDAL/OGR UTILITY PROGRAMS

GDAL/OGR package also contains some command-line utilities for processing geospatial data. The GDAL utilities handle raster data, while OGR utilities process the vector data.

4.8.1. GDAL utilities

The following utility programs are distributed with GDAL.

- gdalinfo- Report information about a file.
- gdal_translate- Copy a raster file, with control of output format.
- gdaladdo- Add overviews to a file.
- gdalwarp- Warp an image into a new coordinate system.
- gdaltindex- Build a map server raster tile index.
- gdalbuildvrt- Build a VRT from a list of datasets.
- gdal_contour- Contours from DEM.
- gdaldem- Tools to analyze and visualize DEMs.
- rgb2pct.py- Convert a 24-bit RGB image to 8-bit paletted.
- pct2rgb.py- Convert an 8-bit paletted image to 24-bit RGB.
- gdal_merge.py- Build a quick mosaic from a set of images.
- gdal2tiles.py- Create a TMS tile structure, KML and simple web viewer.
- gdal_rasterize- Rasterize vector into raster file.
- gdaltransform- Transform coordinates.
- nearblack- Convert nearly black/white borders to exact value.

- gdal_retile.py- Retiles a set of tiles and/or build tiled pyramid levels.
- gdal_grid- Create raster from the scattered data.
- gdal_proximity.py- Compute a raster proximity map.
- gdal_polygonize.py- Generate polygons from raster.
- gdal_sieve.py- Raster Sieve filter.
- gdal_fillnodata.py- Interpolate in no data regions.
- gdallocationinfo- Query raster at a location.
- gdalsrsinfo - Report a given CRS in different formats.
- gdalmove.py - Transform the coordinate system of a file (GDAL >= 2.0).
- gdal-config- Get options required to build software using GDAL.

4.8.2. OGR utilities

Like GDAL, OGR also comes with some powerful utilities.

- ogrinfo- Lists information about an OGR supported data source.
- ogr2ogr- Converts simple features data between file formats.
- ogrtindex- Creates a tile index.

The chapter demonstrated scripts for carrying out various raster and vector data processing tasks.

BIBLIOGRAPHY

- gdal.py, GDAL/OGR 1.9.1 package.
- gdal_array.py, GDAL/OGR 1.9.1 package.
- gdalconst.py, GDAL/OGR 1.9.1 package.
- ogr.py, GDAL/OGR 1.9.1 package.
- osr.py, GDAL/OGR 1.9.1 package.
- Geospatial Data Abstraction Library, *http://www.gdal.org/* [accessed on 31/07/2012].
- Matplotlib- SourceForge Hosting, *http://matplotlib.sourceforge.net/* [accessed on 25/08/2012].
- OSGeo Trac Instances, *http://trac.osgeo.org/* [accessed on 22/08/2012].

Chapter 5
LIDAR DATA PROCESSING

5.1. LIDAR

LIDAR stands for *Light Detection And Ranging*, which is a remote sensing technology for measuring distance and/or other properties of a target using laser pulse. LIDAR technology has a wide variety of applications including archaeology, geology, forestry etc.

5.2. ASPRS

Founded in 1934, the *American Society for Photogrammetry and Remote Sensing* (ASPRS) is an imaging and geospatial information society serving more than 7,000 professionals worldwide. Its mission is to promote the ethical application of active and passive sensors, the disciplines of photogrammetry, remote sensing, geographic information systems, and other supporting geospatial technologies; to advance the understanding of the geospatial and related sciences; to expand public awareness of the profession; and to promote a balanced representation of the interests of government, academia, and private enterprise.

5.3. ASPRS LAS 1.2 FILE FORMAT

The LAS file is intended to contain LIDAR point data records. The data will generally be put into this format from software (e.g. provided by LIDAR hardware vendors) which combines GPS, IMU (Inertial Measurement Unit), and laser pulse range data to produce X, Y, and Z point data. The intention of the data format is to provide an open format that allows different LIDAR hardware and software tools to output data in a common format. The format contains binary data consisting of a header block, variable length records, and point data as shown in table 5-1.

Table 5-1: ASPRS LAS format definition

PUBLIC HEADER BLOCK
VARIABLE LENGTH RECORDS
POINT DATA RECORDS

All data is in little-endian format. The header block consists of a public block followed by variable length records. The public block contains generic data such as point numbers and coordinate bounds. The variable length records contain variable types of data including projection information, metadata, and user application data. The *American Society for Photogrammetry & Remote Sensing* (ASPRS) is the owner of the LAS specification.

5.4. LIBLAS

libLAS is a C/C++ library for reading and writing ASPRS LAS format LIDAR data. It also contains a suite of command-line utilities for inspecting, manipulating, transforming, and processing LAS format LIDAR data. libLAS is available under the terms of BSD license and has APIs for C, C++, .NET, Ruby, and Python.

5.5. INSTALLATION

One can download Microsoft Windows executable file of Python binding of libLAS library (*libLAS-1.6.0.win32.exe*) from *http://pypi.python.org/pypi/libLAS* and install it in *Lib\site-packages* sub-folder of Python installation directory.

5.6. LIBLAS INFORMATION

Move further with exploring some information about libLAS library (result shown in figure 5-1); start with importing libLAS package followed by acquiring libLAS and installed Python version. In this literature, we will be using version 1.6.0 of libLAS library and Python version 2.7.2. Also, complete installation path of libLAS library can be known.

```
import liblas
print 'libLAS version: ',liblas.get_version()
print 'Python version: ',liblas.version
print 'Installation path: ',liblas.__path__
```

By linking *libgeotiff* to libLAS, one can get and set the spatial reference system of LAS file using PROJ.4 coordinate system definitions. Alternatively, by linking libLAS 1.2 (or higher) to GDAL/OGR, one can manipulate spatial references using OGC WKT of ASPRS LAS file. Version 1.6 or greater of libLAS can be used in combination with GDAL/OGR 1.8 (or higher version) to support vertical coordinate systems as well. libLAS 1.6 (or higher) allows you to assign colour information to a LAS file, if GDAL is linked in at compile time. The procedure for compiling/linking GDAL/OGR to libLAS has not been discussed in this book.

```
print 'GDAL enabled/linked: ',liblas.HAVE_GDAL
print 'LibGeoTIFF enabled/linked: ',liblas.HAVE_LIBGEOTIFF
```

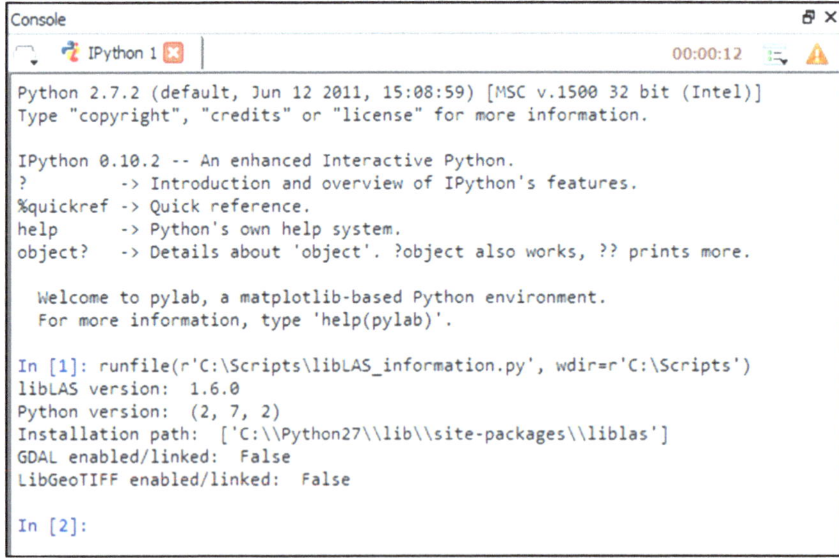

Figure 5-1: libLAS information

5.7. READING LAS HEADER

Move on with writing a script, which gathers some information from header of LAS file. Start with importing libLAS library which is used for accessing ASPRS LAS file. The script was tested with *Barrow_SeaIce_May7_2008.laz*, and the result is shown in figure 5-2. Import libLAS module and open the LAS file using *file.File* class by passing the filename. The *mode* parameter takes the desired file access (default is read mode *r*); while the other modes are write (*w*) and append (*w+*). The *in_srs* parameter accepts the input SRS argument to override the existing file's SRS; while *out_srs* takes the output SRS to reproject points on the fly to, as they are read/written.

```
import liblas
in_file='C:\\Data\\Barrow_SeaIce_May7_2008.laz'
f=liblas.file.File(filename=in_file,mode='r',in_srs=None,out_srs=None)
```

A header contains set of generic data and metadata describing ASPRS LAS file. The header is stored at the beginning of every valid ASPRS LAS file. Fetch the file header so as to explore stored properties.

```
hdr=f.header
```

The file signature must contain the four characters *LASF*, and it is required by the LAS specification. These four characters can be checked by user software as a quick determination of file type.

```
print 'File signature: ',hdr.file_signature
```

File source ID is flight line number, if the file was derived from an original flight line. This field should be set to a value between 1 and 65,535 (inclusive). A value of 0 is interpreted that an ID has not been assigned. In this case, processing software is free to assign any valid number. Note that this scheme allows a LIDAR project to contain up to 65,535 unique sources. A source can be considered an original flight line or it can be the result of merge and/or extract operations.

```
print 'File source ID: ',hdr.filesource_id
```

Project ID (GUID data) consists of four fields comprising of complete Globally Unique Identifier (GUID) are reserved for use as a Project Identifier (Project ID). The field is optional and the time of assignment of the Project ID is at the discretion of processing software. The Project ID should be the same for all files that are associated with a unique project. By assigning a Project ID and using a File source ID (defined above), every file within a project and every point within a file can be uniquely identified, globally.

```
print 'Project ID: ',hdr.project_id
```

The version number consists of a major and minor field. These fields combine to form the number that indicates the format number of the LAS specification. For example, specification number 1.2 would contain 1 in the major field and 2 in the minor field. For all practical purposes, major version number for the file is always 1. libLAS currently supports 0, 1, and 2 as minor versions.

print 'LAS file version: ',hdr.version
print 'Version major: ',hdr.major_version
print 'Version minor: ',hdr.minor_version

LAS version 1.0 specification assumes that files are exclusively generated as a result of collection by a hardware sensor. Subsequent versions recognize that files often result from extraction, merging or modifying existing data files. Thus System ID becomes (as shown in table 5-2):

Table 5-2: System IDs corresponding to LAS file generating agents

Generating Agent	System ID
Hardware system	String identifying hardware (e.g. “ALTM 1210” or “ALS50”)
Merge of one or more files	“MERGE”
Modification of a single file	“MODIFICATION”
Extraction from one or more files	“EXTRACTION”
Reprojection, rescaling etc.	“TRANSFORMATION”
Some other operation	“OTHER” or a string up to 32 characters identifying the operation

print 'System identifier: ',hdr.system_id

Generating software is ASCII data describing the generating software itself. This field provides a mechanism for specifying which generating software package and version was used during LAS file creation (e.g. “TerraScan V-10.8”, “REALM V-4.2” and etc.). If the character data is less than 16 characters, the remaining data must be assigned null.

print 'Generating software: ',hdr.software_id

According to ASPRS LAS specification, header size is the size (in bytes) of the public header block. In the event that the header is extended by a software application through the addition of data at the end of the header, the header size field must be updated with the new header size. Extension of the public header block is discouraged; the variable length records should be used whenever possible to add custom header data. In the event of generating software package adds data to the public header block, this data must be placed at the end of the structure and the header size must be updated to reflect the new size. For libLAS, this is always 227, and it is not configurable.

print 'Header size(in bytes): ',hdr.header_length

Offset to point data is the number of bytes from the beginning of the file to the first field of the first point record data field. This data offset must be updated if any software adds data to the public header block or adds/removes data to/from the variable length records.

print 'Offset to point data (in bytes): ',hdr.data_offset

Number of variable length records field contains the current number of variable length records. This number must be updated if the number of variable length records changes at any time.
print 'Number of Variable Length Records: ',hdr.records_count

Point data format ID corresponds to the point data record format type. LAS 1.2 define types 0, 1, 2 and 3. Data format 0 contains few properties of point data, like X, Y, Z, intensity etc. This format does not contain colour and time information of point data. Point data record format 1 is same as format 0 with addition of GPS time; format 2 is same as format 0 with addition of three colour channels, while format 3 is same as point data record format 2 with addition of GPS time.

print 'Point data format ID: ',hdr.dataformat_id

Number of point records field contains the total number of point records within the file. This value could be different from actual number of points in the file, because of negligence of software.

print 'Number of point records: ',hdr.point_records_count

Number of points by return field contains an array of the total point records per return. The first unsigned long value will be the total number of records from the first return, and the second contains the total number for return two, and so forth up to five returns. libLAS does not manage these values automatically and one has to generate the histogram manually for actual result.

print 'Number of points by return: ',hdr.return_count

X, Y, and Z scale factors fields contain double floating point value that is used to scale the corresponding X, Y, and Z long values within the point records. The corresponding X, Y, and Z scale factor must be multiplied by the X, Y, or Z point record value to get the actual X, Y, or Z coordinate. For example, if the X, Y, and Z coordinates are intended to have two decimal point values, then each scale factor will contain the number 0.01.

print 'Scale factor: ',hdr.scale

X, Y, and Z offset fields should be used to set the overall offset for the point records. In general, these numbers will be 0, however, it should always be used. So to find the actual X coordinate ($X_{coordinate}$) of point record, take the point record X (X_{record}) multiplied by the X scale factor (X_{scale}), and then add the X offset (X_{offset}), as shown in equations are 5-1a to 5-1c.

$$X_{coordinate} = (X_{record} \times X_{scale}) + X_{offset} \quad (5\text{-}1a)$$
$$Y_{coordinate} = (Y_{record} \times Y_{scale}) + Y_{offset} \quad (5\text{-}1b)$$
$$Z_{coordinate} = (Z_{record} \times Z_{scale}) + Z_{offset} \quad (5\text{-}1c)$$

print 'Offset: ',hdr.offset

Maximum and minimum X, Y, Z data are the actual unscaled extents of the LAS point file data, specified in the coordinate system of the LAS data. libLAS does not automatically maintain these values; one needs to calculate these values manually for up-to-date results.

print 'Min: ',hdr.min
print 'Max: ',hdr.max

One can use *compressed* class variable to check whether a LAS file is compressed or not.

print 'Compressed file: ',hdr.compressed

Close the LAS file, after performing the task.

f.close()

```
Console
IPython 1                                                    00:00:11
Python 2.7.2 (default, Jun 12 2011, 15:08:59) [MSC v.1500 32 bit (Intel)]
Type "copyright", "credits" or "license" for more information.

IPython 0.10.2 -- An enhanced Interactive Python.
?         -> Introduction and overview of IPython's features.
%quickref -> Quick reference.
help      -> Python's own help system.
object?   -> Details about 'object'. ?object also works, ?? prints more.

  Welcome to pylab, a matplotlib-based Python environment.
  For more information, type 'help(pylab)'.

In [1]: runfile(r'C:\Scripts\Reading_LAS_header.py', wdir=r'C:\Scripts')
File signature:  LASF
File source ID:  0
Project ID:  00000000-0000-0000-0000-000000000000
LAS file version:  1.1
Version major:  1
Version minor:  1
System identifier:  EXPORT
Generating software:  RiSCANPRO
Header size(in bytes):  227
Offset to point data (in bytes):  413
Number of Variable Length Records:  2
Point data format ID:  1
Number of point records:  2233574
Number of points by return:  [2233574L, 0L, 0L, 0L, 0L, 0L, 0L, 0L]
Scale factor:  [0.0001, 0.0001, 0.0001]
Offset:  [0.0, 0.0, 0.0]
Min:  [-214748.3648, -214748.3648, -2.5353000000000003]
Max:  [-214748.3648, -214748.3648, 86.283]
Compressed file:  True

In [2]:
```

Figure 5-2: Header information of LAS file(Barrow_SeaIce_May7_2008.laz).

5.8. READING LAS POINTS

Now fetch and prints properties of point data in LAS file. Import libLAS library, followed by opening of LAS file (*Barrow_SeaIce_May7_2008.laz*) in read only mode and get the header object.

import liblas
in_file='C:\\Data\\Barrow_SeaIce_May7_2008.laz'

```
f=liblas.file.File(in_file,mode='r')
hdr=f.header
```

The number of point records in the file, according to the header, can also be obtained using *__len__()* method. Use *for* loop to accomplish reading of point data one-by-one.

```
for i in range(f.__len__()):
    print 'Record ',i+1
```

Read the point at the given index using *read()* method.

```
    pnt=f.read(i)
```

X, Y, Z values of point data are stored in *x, y, z* fields as long integers. These are used in conjunction with the scale and offset values to determine the coordinate for each point.

```
    print 'X,Y,Z: ',pnt.x,pnt.y,pnt.z
```

The intensity value is the integer representation of the pulse return magnitude. This value is optional and system specific.

```
    print 'Intensity: ',pnt.intensity
```

Return number is the pulse return number for a given output pulse. A given output laser pulse can have many returns, and they must be marked in sequence of return. The first return will have a return number of 1; the second will be return number of 2, and so on up to five returns.

```
    print 'Return number: ',pnt.return_number
```

Number of returns (for the emitted pulse) is the total number of returns for a given pulse. For example, a laser data point may be return 2 (return number) within a total number of 5 returns.

```
    print 'Number of returns: ',pnt.number_of_returns
```

Scan direction flag denotes the direction at which the scanner mirror was traveling at the time of the output pulse. A bit value of 1 is a positive scan direction, and a bit value of 0 is a negative scan direction (where positive scan direction is a scan moving from the left side of the in-track direction to the right side, and negative the opposite).

```
    print 'Scan direction flag: ',pnt.scan_direction
```

Edge of flight line data bit has a value of 1 only when the point is at the end of a scan. It is the last point on a given scan line before it changes direction.

```
    print 'Edge of flight line: ',pnt.flightline_edge
```

Possible ASPRS standard LIDAR point classes (including classification values) of a given point data are enlisted in table 5-3:

Table 5-3: ASPRS Standard LIDAR Point Classes

Classification value	Meaning
0	Created, never classified
1	Unclassified
2	Ground
3	Low Vegetation
4	Medium Vegetation
5	High Vegetation
6	Building
7	Low Point (noise)
8	Model Key-point (mass point)
9	Water
10	Reserved for ASPRS Definition
11	Reserved for ASPRS Definition
12	Overlap Points
13-31	Reserved for ASPRS Definition

Both 0 and 1 are used for *Unclassified* to maintain compatibility with current popular classification software such as TerraScan. Extend the idea of classification value 1 to include cases in which data have been subjected to a classification algorithm, but emerged in an undefined state. For example, data with class 0 should be sent through an algorithm to detect man-made structures; points that emerge without having been assigned as belonging to structures could be remapped from class 0 to class 1.

```
print 'Classification: ',pnt.classification
```

Scan angle rank is a signed one-byte number with a valid range from -90 to +90. It is the angle (rounded to the nearest integer in the absolute value sense) at which the laser point was output from the laser system including the roll of the aircraft. The scan angle is within 1 degree of accuracy from +90 to –90 degrees. The scan angle is an angle based on 0 degrees being nadir, and –90 degrees to the left side of the aircraft in the direction of flight.

```
print 'Scan angle rank: ',pnt.scan_angle
```

User data field is used at the user's discretion.

```
print 'User data: ',pnt.user_data
```

Point source ID indicates the file from which this point originated. Valid value for this field lies in between 1 to 65,535 (inclusive), with 0 being used for a special case discussed below. The numerical value corresponds to the file source ID from which this point originated. 0 is reserved as a convenience to system implementers. A Point source ID of 0 implies that this point originated in this file. This implies that processing software should set the point source ID equal to the file source ID of the file containing this point at some time during processing.

```
print 'Point source ID: ',pnt.point_source_id
```

Based upon point data format ID, retrieve colour channels (red, green, blue) and/or GPS time information. The GPS time is the double floating point time tag value at which the point was acquired.

```
if hdr.data_format_id==1 or hdr.data_format_id==3:
    print 'GPS time: ',pnt.time
if hdr.data_format_id==2 or hdr.data_format_id==3:
    c=pnt.color
    print 'Red,Green,Blue colors: ',c.red,c.green,c.blue
```

Close the LAS file.

```
f.close()
```

5.9. WRITING LAS FILE

Till now, only reading of header block and point data records of a LAS file has been discussed. Now move a step further with writing a LAS file, in which point data will be stored which has a scan angle of 0. Import libLAS module and get hold of header after opening LAS file (*Barrow_SeaIce_May7_2008.laz*) in read mode.

```
import liblas
in_filename='C:\\Data\\Barrow_SeaIce_May7_2008.laz'
in_file=liblas.file.File(in_filename,mode='r')
in_hdr=in_file.header
```

Also open a new file in write mode and defines a new header object. Assigns the major and minor version of input file to the newly created output file header. The default constructed header is configured according to the ASPRS LAS 1.2 specification, with point data format set to 0 while some other fields are filled with 0.

```
out_filename='C:\\Results\\Output.las'
out_hdr=liblas.header.Header()
out_file=liblas.file.File(out_filename,mode='w',header=out_hdr)
out_hdr.major=in_hdr.major
out_hdr.minor=in_hdr.minor
```

Read all point records one by one from input file, and write only those X, Y, Z values to newly created file which are having scan angle as 0. After that, close all opened files.

```
for p in in_file:
    pnt=liblas.point.Point()
```

```
    if p.scan_angle==0:
        pnt.x=p.x
        pnt.y=p.y
        pnt.z=p.z
        out_file.write(pnt)
in_file.close()
out_file.close()
```

5.10. LAS UTILITY APPLICATIONS

As discussed before, libLAS contains a suite of command-line utilities for inspecting, manipulating, transforming, and processing ASPRS LAS format LIDAR data. These utilities are not enumerated in this literature, but are enlisted below:

- lasinfo
- lasinfo-old
- las2las
- las2las-old
- las2txt
- lasmerge
- lasdiff
- las2ogr
- txt2las
- lasblock
- ts2las
- las2tindex

5.11. DRAWBACKS OF LIBLAS

libLAS has been very successful and is in production use within a number of commercial applications. With respect to the standard (ASPRS LAS specification version 1.3), libLAS does not support the addition of full waveform data. If the application cannot interoperate with C, C++, .NET, Ruby or Python, one needs some other package. One notable problem with libLAS is that it was designed around the point structures defined by the ASPRS LAS standard, and it does not work with other LIDAR point cloud formats, such as ASTM E57 format or the bathymetric BAG format.

BIBLIOGRAPHY

- American Society for Photogrammetry and Remote Sensing, *http://www.asprs.org/* [accessed on 06/03/2012].
- ASPRS LAS Format Version 1.2, *http://www.asprs.org/* [accessed on 01/02/2012].
- libLAS, *http://liblas.org/* [accessed on 08/08/2012].
- LIDAR News, *http://www.lidarnews.com/* [accessed on 19/07/2012].
- Python Programming Language, *http://www.python.org/* [accessed on 10/09/2012].
- Wikipedia, *http://www.wikipedia.org/* [accessed on 05/07/2012].

Chapter 6
GIS DATA IN PUBLIC DOMAIN

Till now, processing of geospatial data has been discussed, without focus on procurement of satellite data or vector data. The motive of this section is to learn about various geospatial data available in public domain which can be downloaded electronically via internet. The following will be discussed:

- ASTER GDEM
- GADM database
- Natural Earth Data
- SRTM

6.1. ASTER GDEM

The *Advanced Spaceborne Thermal Emission and Reflection Radiometer* (ASTER) *Global Digital Elevation Model* (GDEM) is a joint product developed and made available to the public by the Ministry of Economy, Trade, and Industry (METI) of Japan and *National Aeronautics and Space Administration* (NASA) of United States of America. It is generated from data collected from the ASTER, a spaceborne earth observing optical instrument. The ASTER instrument that was launched onboard NASA's Terra spacecraft in December 1999 has an along-track stereoscopic capability using its near infrared spectral band to acquire the stereo data.

A number of 15,14,360 scenes (level-1A products) that was acquired from March, 2000 to August, 2010 was used to generate version 2.0 of ASTER GDEM. It was created by stacking all cloud-masked scene DEMs and non-cloud-masked scene DEMs, and statistical selection algorithm to remove abnormal data.

The ASTER GDEM covers land surfaces between 83°N and 83°S and is composed of 22,702 1°-by-1° tiles. Tiles that contain at least 0.01% land area are included. The ASTER GDEM is in GeoTIFF format with geographic latitude/longitude coordinates and 1 arc-second (30 m) grid of elevation postings. It is referenced to the WGS84/EGM96 geoid. The GDEM has data type as 16-bit signed integer, with elevation information represented as 1m/DN. It also has special DN, -9999 as void pixels and 0 for sea.

Each GDEM tile container incorporates two files, DEM file and Quality Assessment (QA) file, with a dimension of 3601 samples by 3601 lines, correspond to 1°-by-1° data area, and is zip-compressed. The names of individual data tiles refer to the latitude and longitude at the geometric centre of lower-left (southwest) corner pixel. For example, the coordinates of the lower-left corner of the tile *ASTGTM2_N00E006* tile are 0°N latitude and 6°E longitude. *ASTGTM2_N00E006_dem* and *ASTGTM2_N00E006_num* files accommodate DEM and QA data, respectively. Some of the links from which one can electronically download version 2.0 of ASTER GDEM tiles are:

- *http://gdem.ersdac.jspacesystems.or.jp/*
- *http://reverb.echo.nasa.gov/*

6.2. GADM DATABASE

GADM (current version 2.0) is a spatial database of the location of the world's administrative areas (or adminstrative boundaries) for use in GIS software. Administrative areas in this database are countries and lower level subdivisions such as provinces etc. GADM describes where these administrative areas are (the spatial features), and for each area it provides some attributes such as the name etc. This dataset is freely available for academic and other non-commercial use; and can be obtained in the form of shapefile, ESRI geodatabase (MS Access database), RData, and Google Earth KMZ format. The coordinate reference system is latitude/longitude and the WGS84 datum, and the data can be downloaded from the link:*http://www.gadm.org/*.

6.3. NATURAL EARTH DATA

Natural Earth is a public domain (free to use for personal, academic or commercial purpose) map dataset available at 1:10 million, 1:50 million, and 1:110 million scales. Natural Earth Vector comes in ESRI shapefile format while raster comes in TIFF format with a world file (TFW). All Natural Earth data uses WGS84 geographic coordinate system. The data is classified in three themes, namely, cultural vector data, physical vector data, and raster data. The cultural vector data theme includes country boundaries, disputed areas, first order administrative boundaries (provinces, states, etc.), populated places (includes capitals, major cities and towns) etc. Physical vector data theme includes data for ocean coastline, land, island, rivers and lakes centrelines, glaciated areas, geographic lines (polar circles, tropical circles, equator) etc. Raster data theme includes shaded relief of land derived from SRTM data, land cover data etc. The natural earth data can be downloaded from the link: *http://www.naturalearthdata.com/*.

6.4. SRTM

The *Shuttle Radar Topography Mission* (SRTM) was flown aboard the space shuttle *Endeavour* on February 11-22, 2000. NASA and the *National Geospatial-Intelligence Agency* (NGA) participated in this international project to acquire radar data which was used to create detailed topographic maps. SRTM data is intended for scientific use with GIS or other special application software. *Endeavour* orbited earth 16 times each day during the 11 day mission completing 176 orbits. SRTM successfully collected radar data over 80% of the Earth's land surface between 60°N and 56°S latitude with data points posted every 1 arc-second (approximately 30 metre).

EarthExplorer (*http://earthexplorer.usgs.gov/*) provides two resolutions of finished grade SRTM data; 1 arc-second (approximately 30-metre) high resolution elevation data is available only for the United States of America, while 3 arc-second (approximately 90-metre) medium resolution elevation data is available for global coverage. The 3 arc-second data was resampled using cubic convolution interpolation for regions between 60°N and 56°S latitude. The data has WGS84 as horizontal datum while vertical datum is EGM96 (Earth Gravitational Model 1996) with vertical units as metre. This elevation data is available in two file formats:

- *Digital Terrain Elevation Data* (DTED) is a standard mapping format with a regularly spaced grid of elevation points designed by the NGA, measured in geographic latitude and longitude units.

- *Band interleaved by line* (BIL) is a binary raster format with an accompanying header file which describes the layout and formatting of the file. This format is recommended for software packages that do not support the DTED format.

CGIAR-CSI portal (*http://srtm.csi.cgiar.org/*) also provides SRTM 3 arc-second DEM. The *Consultative Group on International Agricultural Research (CGIAR)* is a global partnership dedicated for reducing poverty and hunger, improving human health and nutrition, and enhancing ecosystem resilience through advanced agricultural research. The *Consortium for Spatial Information (CGIAR-CSI)* is the CGIAR community that promotes and practices the application of spatial science in achieving these goals effectively.

The SRTM 3 arc-second DEM's have a resolution of 90 metre at the equator, and are provided in mosaiced 5 degree x 5 degree tiles in geographic coordinate system- WGS84 datum for easy download and use. These are available in both ArcInfo ASCII and GeoTIFF format to facilitate their ease of use in a variety of image processing and GIS applications. In addition, a binary data mask file is available for download, allowing user to identify the areas within each DEM which has been interpolated. Some of the links from which one can download SRTM data are:

- *http://earthexplorer.usgs.gov/*
- *http://srtm.csi.cgiar.org/*

BIBLIOGRAPHY

- CGIAR-CSI SRTM database, *http://srtm.csi.cgiar.org/* [accessed on 01/07/2012].
- Earth Explorer, *http://earthexplorer.usgs.gov/* [accessed on 08/07/2012].
- Earth Resources Observation and Science Centre, *http://eros.usgs.gov/* [accessed on 04/09/2012].
- Global Administrative Areas, *http://www.gadm.org/* [accessed on 14/07/2012].
- Japan Space Systems, *http://www.jspacesystems.or.jp/ersdac/* [accessed on 11/09/2012].
- Natural Earth, *http://www.naturalearthdata.com/* [accessed on 22/08/2012].
- Reverb, *http://reverb.echo.nasa.gov/* [accessed on 25/05/2012].

APPENDIX

This section provides all the scripts that are discussed in this book.

PYPROJ INFORMATION

```
#!/usr/bin/env python
import pyproj
print 'Pyproj package version: ',pyproj.__version__
print 'Pyproj package installation path: ',pyproj.__path__
print 'PROJ.4 data path: ',pyproj.pyproj_datadir
print '\n\nSupported map projections (code::map projection name):'
for list in pyproj.pj_list:
   print '{0}::{1}'.format(list,pyproj.pj_list[list])
print '\n\nSupported ellipsoids (ellipsoid code::ellipsoid details):'
for list in pyproj.pj_ellps:
   print '{0}::{1}'.format(list,pyproj.pj_ellps[list])
```

COORDINATE TRANSFORMATION

```
#!/usr/bin/env python
import pyproj
proj_wgs84=pyproj.Proj(init='epsg:4326',preserve_units=True)
#proj_wgs84=pyproj.Proj(init='esri:4326',preserve_units=True)
#proj_wgs84=pyproj.Proj(projparams="+init=epsg:4326",preserve_units=True)
#proj_wgs84=pyproj.Proj(projparams="+init=esri:4326",preserve_units=True)
print 'PROJ version:',proj_wgs84.proj_version
if proj_wgs84.is_latlong():
   print "EPSG:4326 coordinates are in lat/lon"
else:
   print "EPSG:4326 coordinates are not in lat/lon"
proj_wgs84_33s=pyproj.Proj(projparams='''+proj=utm +zone=33 +south +datum=WGS84
                  +units=m''',preserve_units=True)
lat=-8.852400;lon=13.237200
x1,y1=pyproj.transform(p1=proj_wgs84,p2=proj_wgs84_33s,x=lon,y=lat,z=None, \
                radians=False)
print 'Projected coordinates- X: %9.9f, Y: %9.9f'%(x1,y1)
proj_wgs84_33s=pyproj.Proj(proj='utm',zone=33,south=True,datum='WGS84', \
                  preserve_units=True)
#proj_wgs84_33s=pyproj.Proj(projparams={'proj':'utm','zone':33,'south':'True', \
#                  'datum':'WGS84','units':'m'},preserve_units=True)
if proj_wgs84_33s.is_geocent():
   print "WGS 84 UTM 33S is geocentric coordinate system"
else:
   print "WGS 84 UTM 33S is not geocentric coordinate system"
lon,lat=proj_wgs84_33s(x1,y1,inverse=True,radians=False)
print 'Geographic coordinates- Longitude: %9.9f, Latitude: %9.9f' %(lon,lat)
proj_wgs84=proj_wgs84_44n=None
```

GEODETIC COMPUTATION

```
#!/usr/bin/env python
from pyproj import Geod
g=Geod(initstring='+ellps=clrk66')
#g=Geod(initstring='+a=6378206.4 +b=6356583.8')
#g=Geod(ellps='clrk66')
#g=Geod(a=6378206.4,b=6356583.8)
boston_lat=42.+(15./60.); boston_lon=-71.-(7./60.)
portland_lat=45.+(31./60.); portland_lon=-123.-(41./60.)
az12,az21,dist=g.inv(lons1=boston_lon,lats1=boston_lat,lons2=portland_lon, \
            lats2=portland_lat,radians=False)
print "Forward azimuth: %7.3f" %(az12)
print "Backward azimuth: %7.3f" %(az21)
print "Distance: %12.3f" % (dist)
endlon,endlat,backaz=g.fwd(lons=boston_lon,lats=boston_lat,az=az12,dist=dist, \
               radians=False)
print "Portland latitude: %7.3f" %(endlat)
print "Portland longitude: %7.3f" %(endlon)
print "Backward azimuth: %7.3f" % (backaz)
lonlats=g.npts(lon1=boston_lon,lat1=boston_lat,lon2=portland_lon, \
         lat2=portland_lat,npts=10,radians=False)
print 'Ten equally spaced points between Boston and Portland (lat, lon):'
for lon,lat in lonlats: print '%6.3f  %7.3f' %(lat,lon)
g=None
```

SRS INFORMATION

```
#!/usr/bin/env python
import osgeo.osr as osr
print "Supported map projections:"
for pm in osr.GetProjectionMethods():
   print pm[1]
SRS_GCS=osr.SpatialReference()
SRS_GCS.ImportFromEPSG(4326)
SRS_PCS=osr.SpatialReference()
SRS_PCS.ImportFromEPSG(32643)
xform=osr.CoordinateTransformation(SRS_GCS,SRS_PCS)
print "Coordinate transformation from GCS to PCS"
print xform.TransformPoint(75.867937,22.699574,0)
print "SRS_PCS description:"
if SRS_PCS.IsGeographic():
   print "SRS_PCS is a Geographic coordinate system."
if SRS_PCS.IsProjected():
   print "SRS_PCS is a Projected coordinate system."
if SRS_PCS.IsGeocentric():
   print "SRS_PCS is a Geocentric coordinate system."
if SRS_PCS.IsCompound():
   print "SRS_PCS is a Compound coordinate system."
if SRS_PCS.IsLocal():
   print "SRS_PCS is a Local coordinate system."
if SRS_PCS.IsVertical():
   print "SRS_PCS is a Vertical coordinate system."
print "UTM Zone: ",SRS_PCS.GetUTMZone()
print "Projected coordinate system: ",SRS_PCS.GetAttrValue("PROJCS",0)
print "Geographic coordinate system: ",SRS_PCS.GetAttrValue("GEOGCS",0)
print "Projection: ",SRS_PCS.GetAttrValue("PROJECTION",0)
print "Datum: ",SRS_PCS.GetAttrValue("DATUM",0)
print "Spheroid: ",SRS_PCS.GetAttrValue("SPHEROID",0)
print "Semi major axis: ",SRS_PCS.GetSemiMajor()
print "Semi minor axis: ",SRS_PCS.GetSemiMinor()
print "Inverse flattening: ",SRS_PCS.GetInvFlattening()
print "Prime Meridian: ",SRS_PCS.GetAttrValue("PRIMEM",0)
print "Projection parameter (Latitude of origin): ",SRS_PCS.GetProjParm("latitude_of_origin")
print "Projection parameter (Central meridian): ",SRS_PCS.GetProjParm("central_meridian")
print "Projection parameter (Scale factor): ",SRS_PCS.GetProjParm("scale_factor")
print "Projection parameter (False easting): ",SRS_PCS.GetProjParm("false_easting")
print "Projection parameter (False northing): ",SRS_PCS.GetProjParm("false_northing")
print "Linear unit name: ",SRS_PCS.GetLinearUnitsName()
print "Linear unit: ",SRS_PCS.GetLinearUnits()
```

```
print "Angular unit name: ",SRS_PCS.GetAttrValue("GEOGCS|UNIT",0)
print "Angular unit: ",SRS_PCS.GetAngularUnits()
print "Authority name: ",SRS_PCS.GetAuthorityName("PROJCS")
print "Authority code: ",SRS_PCS.GetAuthorityCode("PROJCS")
print "WKT format: \n",SRS_PCS.ExportToWkt()
print "Pretty WKT format: \n",SRS_PCS.ExportToPrettyWkt(0)
print "Proj.4 format: \n",SRS_PCS.ExportToProj4()
SRS_PCS.MorphToESRI()
print "ESRI compatible WKT for use as *.prj: \n", SRS_PCS.ExportToWkt()
print "MapInfo format: \n",SRS_PCS.ExportToMICoordSys()
SRS_GCS=SRS_PCS=None
```

BASEMAP INFORMATION

```
#!/usr/bin/env python
import mpl_toolkits.basemap as basemap
print "Matplotlib version: ",basemap._matplotlib_version
print "Matplotlib basemap toolkit version: ",basemap.__version__
print "Matplotlib basemap toolkit data directory: ",basemap.basemap_datadir
print "\nSupported map projections:\n",basemap.supported_projections
print 'Supported map projection parameters (code:projection parameters):'
for prj_param in basemap.projection_params:
   print '{0}\t: {1}'.format(prj_param,basemap.projection_params[prj_param])
print "\nCylindrical projections: ",basemap._cylproj
print "Pseudocylindrical projections: ",basemap._pseudocyl
```

SUPPORTED COLOURS

```
#!/usr/bin/env python
import matplotlib.colors as colors
print "Basic built-in colours (code:RGB tuple)"
for clr in colors.ColorConverter.colors:
   print '{0}: {1}'.format(clr,colors.ColorConverter.colors[clr])
print "\nAll supported colours (colour:Hex string:RGB tuple)"
for clr in colors.cnames:
   HEX=colors.cnames[clr]
   print '{0}:{1}:{2}'.format(clr,HEX,colors.hex2color(HEX))
```

MAP MAKING

```
#!/usr/bin/env python
from mpl_toolkits.basemap import Basemap
from mpl_toolkits.axes_grid1.inset_locator import inset_axes
import matplotlib.pyplot as plt
from matplotlib.patches import Polygon
import numpy as np
fig=plt.figure(figsize=(15,15),facecolor='0.8')
fig.canvas.set_window_title(title="Cartography")
fig.suptitle(t='Maps',x=0.5,y=0.5,color='r',fontsize=22,fontweight='bold', \
        fontstyle='italic')
ax1=fig.add_subplot(221)
bmap1=Basemap(projection='tmerc',lon_0=82,lat_0=22,k_0=0.9996, \
        rsphere=(6378137.00,6356752.314245179),resolution='l', \
        width=3000000,height=3500000,ax=ax1)
bmap1.etopo(scale=0.5,ax=ax1)
ax1.set_title(label="ETOPO Relief Map",color='k',fontsize=14)
ax2=fig.add_subplot(222)
bmap2=Basemap(projection='tmerc',lon_0=-90,lat_0=40,k_0=0.9996, \
        rsphere=(6378137.00,6356752.314245179),resolution='c', \
        width=6000000,height=6000000,ax=ax2)
bmap2.drawmapboundary(color='k',linewidth=1.0,fill_color='aqua',ax=ax2)
bmap2.drawcoastlines(linewidth=1.0,color='k',antialiased=1,ax=ax2)
bmap2.fillcontinents(color='coral',lake_color='aqua',ax=ax2)
bmap2.drawcountries(linewidth=1.0,color='k',antialiased=1,ax=ax2)
bmap2.drawstates(linewidth=0.5,color='g',antialiased=1,ax=ax2)
ax2.set_title(label="Thematic Map",color='k',fontsize=14)
axin=inset_axes(parent_axes=bmap2.ax,width="30%",height="30%",loc=4)
omap=Basemap(projection='ortho',lon_0=-105,lat_0=40,ax=axin)
omap.drawcountries(color='white',ax=axin)
omap.fillcontinents(color='gray',ax=axin)
bx,by=omap(bmap2.boundarylons,bmap2.boundarylats)
xy=zip(bx,by)
mapboundary=Polygon(xy,edgecolor='red',linewidth=2,fill=False)
omap.ax.add_patch(mapboundary)
ax3=fig.add_subplot(223)
bmap3=Basemap(projection='tmerc',lon_0=82,lat_0=22,k_0=0.9996, \
        rsphere=(6378137.00,6356752.314245179),resolution=None, \
        width=3000000,height=3500000,ax=ax3)
bmap3.shadedrelief(scale=0.35,ax=ax3)
x1,y1=0.75*bmap3.xmax,0.1*bmap3.ymax
lon1,lat1=bmap3(x1,y1,inverse=True)
bmap3.drawmapscale(lon=lon1,lat=lat1,lon0=lon1,lat0=lat1,length=1000, \
```

```
            barstyle='fancy',units='km',fontsize=9,yoffset=None, \
            labelstyle='simple',fontcolor='k',fillcolor1='w', \
            fillcolor2='k',format='%d',ax=ax3)
ax3.set_title(label="Shaded Relief Map",color='k',fontsize=14)
ax4=fig.add_subplot(224)
bmap4=Basemap(projection='tmerc',lon_0=95,lat_0=22,k_0=0.9996, \
        rsphere=(6378137.00,6356752.314245179),resolution='c', \
        width=9500000,height=7500000,ax=ax4)
meridians=np.arange(0,360,10)
bmap4.drawmeridians(meridians,color='y',linewidth=1.0,dashes=[1,1], \
            labels=[0,0,0,1],labelstyle=None,xoffset=None,yoffset=None, \
            latmax=None,ax=ax4)
circles=np.arange(-90,90,10)
bmap4.drawparallels(circles,color='y',linewidth=1.0,dashes=[1,1], \
            labels=[1,0,0,0],labelstyle=None,xoffset=None,yoffset=None, \
            latmax=None,ax=ax4)
lat=[28.63,31.23]
lon=[77.22,121.47]
cities=['New Delhi','Shanghai']
bmap4.drawgreatcircle(lon1=lon[0],lat1=lat[0],lon2=lon[1],lat2=lat[1], \
             linewidth=2,color='m',ax=ax4)
x,y=bmap4(lon,lat)
plt.plot(x,y,'ro')
for city,xpt,ypt in zip(cities,x,y):
   plt.text(x=xpt-1000000,y=ypt+200000,s=city,bbox=dict(facecolor='yellow', \
        alpha=0.5))
bmap4.drawmapscale(lon=110,lat=-5,lon0=110,lat0=-5,length=2000, \
            barstyle='simple',units='km',fontsize=9,yoffset=None, \
            labelstyle='simple',fontcolor='r',format='%d',ax=ax4)
bmap4.bluemarble(scale=0.5,ax=ax4)
ax4.set_title(label="Blue Marble Map",color='k',fontsize=14)
fname='C:/Results/Maps.png'
plt.savefig(fname,dpi=300,facecolor='0.8',bbox_inches='tight',pad_inches=0.5)
fig=bmap1=bmap2=bmap3=bmap4=None
```

RASTER INFORMATION

```
#!/usr/bin/env python
try:
   import osgeo.gdal as gdal
   import osgeo.gdal_array as gdal_array
   import osgeo.osr as osr
except ImportError:
   import gdal
   import gdal_array
   import osr
import matplotlib.pyplot as plt
import numpy as np
import sys,os
gdal.AllRegister()
file='C:/Data/small_world.tiff'
dataset=gdal.Open(file,gdal.GA_ReadOnly)
if dataset is None:
   print 'Unable to open file'
   sys.exit(1)
AllRasterData=gdal_array.DatasetReadAsArray(ds=dataset,xoff=0,yoff=0, \
                              xsize=None,ysize=None)
print("Driver: %s/%s" %(dataset.GetDriver().ShortName, \
                  dataset.GetDriver().LongName))
FileList=dataset.GetFileList()
if FileList is None or len(FileList)==0:
   print("Files: none associated")
else:
   print("Files: %s" %FileList[0])
   for i in range(1, len(FileList)):
      print("      %s" %FileList[i])
print("Size is %dx%dx%d" %(dataset.RasterYSize,dataset.RasterXSize, \
                              dataset.RasterCount))
Projection=dataset.GetProjectionRef()
if Projection is not '':
   SRS=osr.SpatialReference()
   if SRS.ImportFromWkt(Projection) is 0:
      PrettyWkt=SRS.ExportToPrettyWkt(False)
      print("Coordinate System is:\n%s" %PrettyWkt)
   else:
      print("Coordinate System is:\n%s" %Projection)
geotrans=dataset.GetGeoTransform()
if geotrans is not None:
   print 'Geotransformation parameters:'
```

```
    print 'Origin(Top left X): ',geotrans[0]
    print 'Origin(Top left Y): ',geotrans[3]
    print 'Pixel size in X direction: ',geotrans[1]
    print 'Pixel size in Y direction: ',geotrans[5]
    print 'Rotation: ',geotrans[2]
    print 'Rotation: ',geotrans[4]
if dataset.GetGCPCount()>0:
    Projection=dataset.GetGCPProjection()
    if Projection is not None:
        SRS=osr.SpatialReference()
        if SRS.ImportFromWkt(Projection)==0:
            PrettyWkt=SRS.ExportToPrettyWkt(False)
            print("GCP Projection: \n%s" %PrettyWkt)
        else:
            print("GCP Projection: %s" %Projection)
    gcps=dataset.GetGCPs()
    i=0
    for gcp in gcps:
        print("GCP[%3d]: Id=%s, Info=%s\n" \
        "       (%.15g,%.15g)->(%.15g,%.15g,%.15g)" \
              %(i,gcp.Id,gcp.Info,gcp.GCPPixel,gcp.GCPLine, \
              gcp.GCPX,gcp.GCPY,gcp.GCPZ))
        i = i + 1
Metadata=dataset.GetMetadata_List("")
if Metadata is not None and len(Metadata)>0:
    print("Default domain metadata:")
    for md in Metadata:
        print("%s" %md)
Metadata=dataset.GetMetadata_List("IMAGE_STRUCTURE")
if Metadata is not None and len(Metadata)>0:
    print("IMAGE_STRUCTURE domain metadata:")
    for md in Metadata:
        print("%s" %md)
Metadata=dataset.GetMetadata_List("SUBDATASETS")
if Metadata is not None and len(Metadata)>0:
    print("SUBDATASETS domain metadata:")
    for md in Metadata:
        print("%s" %md)
Metadata=dataset.GetMetadata_List("GEOLOCATION")
if Metadata is not None and len(Metadata)>0:
    print("GEOLOCATION domain metadata:")
    for md in Metadata:
        print("%s" %md)
Metadata=dataset.GetMetadata_List("RPC")
```

```
if Metadata is not None and len(Metadata)>0:
   print("RPC domain metadata:")
   for md in Metadata:
      print("%s" %md)
print 'Corner coordinates'
print("Upper left: %.5f,%.5f" %(geotrans[0],geotrans[3]))
print("Upper right: %.5f,%.5f" %(geotrans[0]+geotrans[1]*dataset.RasterXSize, \
                        geotrans[3]))
print("Lower left: %.5f,%.5f" %(geotrans[0], \
 geotrans[3]+geotrans[5]*dataset.RasterYSize))
print("Lower right: %.5f,%.5f" %(geotrans[0]+geotrans[1]*dataset.RasterXSize, \
                        geotrans[3]+geotrans[5]*dataset.RasterYSize))
for iBand in range(dataset.RasterCount):
   Band=dataset.GetRasterBand(iBand+1)
   print('Band %d information:' %(iBand+1))
   BlockXSize,BlockYSize=Band.GetBlockSize()
   print("Block size: %dx%d" %(BlockYSize,BlockXSize))
   print("Data type: %s" %(gdal.GetDataTypeName(Band.DataType)))
   if Band.GetDescription() is not None and len(Band.GetDescription())>0:
      print("Description: %s" %Band.GetDescription())
   min=Band.GetMinimum()
   max=Band.GetMaximum()
   if min is None or max is None:
      line=""
      if min is not None:
         line=line+("Min: %.3f " %min)
      if max is not None:
         line=line+("Max: %.3f " %max)
      MinMax=Band.ComputeRasterMinMax(False)
      line=line+("Computed Min/Max: %.3f,%.3f" \
                %(MinMax[0],MinMax[1]))
      print(line)
   print("Checksum: %d" %Band.Checksum(xoff=0, \
          yoff=0,xsize=None,ysize=None))
   NoData=Band.GetNoDataValue()
   if NoData is not None:
      print("NoData value: %.18g" %NoData)
   if Band.GetOverviewCount()>0:
      line="Overviews: "
      for iOverview in range(Band.GetOverviewCount()):
         if iOverview!=0:
            line=line+", "
         Overview=Band.GetOverview(iOverview)
         if Overview is not None:
```

```
            line=line+("%dx%d" %(Overview.XSize,Overview.YSize))
        else:
            line=line+"(null)"
      print(line)
    if Band.HasArbitraryOverviews():
      print("Overviews: arbitrary")
    if len(Band.GetUnitType())>0:
      print("Unit Type: %s" %Band.GetUnitType())
    if Band.GetScale()!=1.0 or Band.GetOffset()!=0.0:
      print("Offset: %.15g, Scale: %.15g" % \
            (Band.GetOffset(),Band.GetScale()))
    Metadata=Band.GetMetadata_List("")
    if Metadata is not None and len(Metadata)>0:
      print("Default domain metadata:")
      for md in Metadata:
        print("%s" %md)
    Metadata=Band.GetMetadata_List("IMAGE_STRUCTURE")
    if Metadata is not None and len(Metadata)>0:
      print("IMAGE_STRUCTURE domain metadata:")
      for md in Metadata:
        print("%s" %md)
    print("Colour interpretation: %s" \
    %(gdal.GetColorInterpretationName(Band.GetRasterColorInterpretation())))
    CTable=Band.GetRasterColorTable()
    if Band.GetRasterColorInterpretation()==gdal.GCI_PaletteIndex \
                              and CTable is not None:
      print("Colour Table (%s with %d entries)" % \
      (gdal.GetPaletteInterpretationName( \
      CTable.GetPaletteInterpretation()), \
      CTable.GetCount()))
      for i in range(CTable.GetCount()):
        Entry=CTable.GetColorEntry(i)
        print( "  %3d: %d,%d,%d,%d" % ( \
        i,Entry[0],Entry[1],Entry[2],Entry[3]))
AllRasterData=np.rollaxis(AllRasterData,0,3)
plt.imshow(X=AllRasterData)
plt.axis("off")
(filepath,filename)=os.path.split(file)
plt.title(filename)
dataset=None
```

NDVI PREPARATION

```
#!/usr/bin/env python
try:
    import osgeo.gdal as gdal
    import osgeo.gdalconst as gdalconst
except ImportError:
    import gdal
    import gdalconst
import sys,numpy
gdal.AllRegister()
File_R_Band='C:/Data/p145r045_7t20011020_z43_nn30.tif'
Dataset_R_Band=gdal.Open(File_R_Band,gdalconst.GA_ReadOnly)
File_NIR_Band='C:/Data/p145r045_7t20011020_z43_nn40.tif'
Dataset_NIR_Band=gdal.Open(File_NIR_Band,gdalconst.GA_ReadOnly)
if Dataset_NIR_Band is None or Dataset_R_Band is None:
    print 'Unable to open file'
    sys.exit(1)
rows_R_Band=Dataset_R_Band.RasterYSize
cols_R_Band=Dataset_R_Band.RasterXSize
Band_R=Dataset_R_Band.GetRasterBand(1)
NoDataVal_R=Band_R.GetNoDataValue()
Band_NIR=Dataset_NIR_Band.GetRasterBand(1)
NoDataVal_NIR=Band_NIR.GetNoDataValue()
driver=Dataset_R_Band.GetDriver()
format=driver.ShortName
metadata=driver.GetMetadata()
if metadata.has_key(gdal.DCAP_CREATE) \
   and metadata[gdal.DCAP_CREATE]=='YES':
      print 'Driver %s supports Create() method.' %format
if metadata.has_key(gdal.DCAP_CREATECOPY) \
   and metadata[gdal.DCAP_CREATECOPY]=='YES':
      print 'Driver %s supports CreateCopy() method.' %format
File_NDVI='C:/Results/NDVI.tif'
outDataSet=driver.Create(utf8_path=File_NDVI,xsize=cols_R_Band, \
                 ysize=rows_R_Band,bands=1,eType=gdalconst.GDT_Float32)
outDataSet.SetGeoTransform(Dataset_R_Band.GetGeoTransform())
outDataSet.SetProjection(Dataset_R_Band.GetProjection())
outBand=outDataSet.GetRasterBand(1)
outBand.SetNoDataValue(-999)
ndvi=numpy.empty((1,cols_R_Band),numpy.float32)
for row in range(rows_R_Band):
    Data_R_Band=Band_R.ReadAsArray(xoff=0,yoff=row,win_xsize=cols_R_Band, \
                      win_ysize=1).astype(numpy.float32)
```

```
    Data_NIR_Band=Band_NIR.ReadAsArray(xoff=0,yoff=row,win_xsize=cols_R_Band, \
                          win_ysize=1).astype(numpy.float32)
    condlist=[Data_R_Band==NoDataVal_R,Data_NIR_Band==NoDataVal_NIR, \
                          (Data_R_Band+Data_NIR_Band)==0]
    choicelist=[-999,-999,-999]
    ndvi=numpy.select(condlist,choicelist, \
    ((Data_NIR_Band-Data_R_Band)/(Data_NIR_Band+Data_R_Band)))
    outBand.WriteArray(array=ndvi,xoff=0,yoff=row)
Dataset_R_Band=Dataset_NIR_Band=outDataSet=None
```

REPROJECTION AND RESAMPLING

```
#!/usr/bin/env python
import osgeo.gdal as gdal
import osgeo.gdal_array as gdal_array
import osgeo.osr as osr
import matplotlib.pyplot as plt
import math,numpy
driver=gdal.GetDriverByName('GTiff')
driver.Register()
inFile='C:/Data/SatImage.tif'
outFile='C:/Results/ReprojectedImg.tif'
inDataset=gdal.Open(inFile,0)
inRasterData=gdal_array.DatasetReadAsArray(inDataset)
in_srs=osr.SpatialReference()
in_srs.ImportFromWkt(inDataset.GetProjection())
in_DataType=inDataset.GetRasterBand(1).DataType
in_col=inDataset.RasterXSize
in_row=inDataset.RasterYSize
in_band=inDataset.RasterCount
in_geo=inDataset.GetGeoTransform()
in_pixelX,in_pixelY=in_geo[1],-in_geo[5]
in_ulx,in_uly=in_geo[0],in_geo[3]
in_urx,in_ury=in_ulx+in_pixelX*in_col,in_uly
in_lrx,in_lry=in_urx,in_uly-in_pixelY*in_row
in_llx,in_lly=in_ulx,in_lry
out_srs=osr.SpatialReference()
out_srs.ImportFromEPSG(32616)
tx=osr.CoordinateTransformation(in_srs,out_srs)
(out_ulx,out_uly,out_ulz)=tx.TransformPoint(in_ulx,in_uly)
(out_urx,out_ury,out_urz)=tx.TransformPoint(in_urx,in_ury)
(out_lrx,out_lry,out_lrz)=tx.TransformPoint(in_lrx,in_lry)
(out_llx,out_lly,out_llz)=tx.TransformPoint(in_llx,in_lly)
minX=min(out_ulx,out_llx)
maxX=max(out_urx,out_lrx)
minY=min(out_lly,out_lry)
maxY=max(out_ury,out_uly)
opt=raw_input('''Enter 1 for specifying rows/columns of reprojected raster,
        2 for specifying pixel resolution of reprojected raster: ''')
if int(opt)==1:
  out_row=int(raw_input('Enter number of rows of reprojected raster: '))
  out_col=int(raw_input('Enter number of columns of reprojected raster: '))
elif int(opt)==2:
  out_pixelX=float(raw_input('Enter pixel width of reprojected raster: '))
```

```
   out_pixelY=float(raw_input('Enter pixel height of reprojected raster: '))
   out_col=int(math.floor((maxX-minX)/out_pixelX))
   out_row=int(math.floor((maxY-minY)/out_pixelY))
out_pixelX=(maxX-minX)/out_col
out_pixelY=(maxY-minY)/out_row
out_geo=(minX,out_pixelX,0,maxY,0,-out_pixelY)
mem_drv=gdal.GetDriverByName('MEM')
outDataset=mem_drv.Create('',out_col,out_row,in_band,in_DataType)
outDataset.SetGeoTransform(out_geo)
outDataset.SetProjection(out_srs.ExportToWkt())
for iBand in range(inDataset.RasterCount):
   in_band=inDataset.GetRasterBand(iBand+1)
   NoDataVal=in_band.GetNoDataValue()
   outDataset.GetRasterBand(iBand+1).SetNoDataValue(NoDataVal)
gdal.ReprojectImage(inDataset,outDataset,in_srs.ExportToWkt(), \
            out_srs.ExportToWkt(),gdal.GRA_NearestNeighbour)
dst_ds=driver.CreateCopy(utf8_path=outFile,src=outDataset,strict=0)
inRasterData=numpy.rollaxis(inRasterData,0,3)
fig1=plt.figure(figsize=(9,7),facecolor='white')
fig1.canvas.set_window_title("Original image")
ax1=fig1.add_subplot(1,1,1)
cax=ax1.imshow(X=inRasterData)
ax1.axis("off")
outRasterData=gdal_array.DatasetReadAsArray(outDataset)
outRasterData=numpy.rollaxis(outRasterData,0,3)
fig2=plt.figure(figsize=(9,7),facecolor='white')
fig2.canvas.set_window_title("Reprojected image")
ax2=fig2.add_subplot(1,1,1)
cax=ax2.imshow(X=outRasterData)
ax2.axis("off")
inDataset=outDataset=dst_ds=fig1=fig2=None
Dataset=gdal.Open(outFile,gdal.GA_ReadOnly)
Dataset.BuildOverviews(resampling="NEAREST",overviewlist=[2,4,8,16,32])
Dataset=None
```

VECTOR INFORMATION

```
#!/usr/bin/env python
try:
    import osgeo.ogr as ogr
except ImportError:
    import ogr
from mpl_toolkits.basemap import Basemap
import matplotlib.pyplot as plt
import sys,os

def FeatureInfo(Feature):
    feature_Defn=Feature.GetDefnRef()
    print("OGRFeature(%s): %ld" %(feature_Defn.GetName(),Feature.GetFID()))
    for iField in range(feature_Defn.GetFieldCount()):
        field_Defn=feature_Defn.GetFieldDefn(iField)
        line="  %s (%s) : " %(field_Defn.GetNameRef(), \
                ogr.GetFieldTypeName(field_Defn.GetType()))
        if Feature.IsFieldSet(iField):
            line=line+"%s" %(Feature.GetFieldAsString(iField))
        else:
            line=line+"(null)"
        print(line)
    if Feature.GetStyleString() is not None:
        print("Style: %s" %Feature.GetStyleString())
    Geometry=Feature.GetGeometryRef()
    if Geometry is not None:
        GeometryInfo(Geometry, "  ")
    print('')
    return

def GeometryInfo(Geometry,Prefix):
    if Prefix == None:
        Prefix = ""
    line=("%s%s : " %(Prefix,Geometry.GetGeometryName()))
    eType=Geometry.GetGeometryType()
    if eType==ogr.wkbLineString or eType==ogr.wkbLineString25D:
        line=line+("%d points" %Geometry.GetPointCount())
        print(line)
    elif eType==ogr.wkbPolygon or eType==ogr.wkbPolygon25D:
        nRings=Geometry.GetGeometryCount()
        if nRings==0:
            line=line+"empty"
        else:
```

```
            poRing=Geometry.GetGeometryRef(0)
            line=line+("%d points" %poRing.GetPointCount())
            if nRings>1:
                line=line+(", %d inner rings (" %(nRings-1))
                for ir in range(0,nRings-1):
                    if ir>0:
                        line=line+", "
                    poRing=Geometry.GetGeometryRef(ir+1)
                    line=line+("%d points" %poRing.GetPointCount())
                line=line+")"
        print(line)
    elif eType==ogr.wkbMultiPoint or \
         eType==ogr.wkbMultiPoint25D or \
         eType==ogr.wkbMultiLineString or \
         eType==ogr.wkbMultiLineString25D or \
         eType==ogr.wkbMultiPolygon or \
         eType==ogr.wkbMultiPolygon25D or \
         eType==ogr.wkbGeometryCollection or \
         eType==ogr.wkbGeometryCollection25D:
            line=line+"%d geometries:" %Geometry.GetGeometryCount()
            print(line)
            for ig in range(Geometry.GetGeometryCount()):
                subgeom=Geometry.GetGeometryRef(ig)
                from sys import version_info
                if version_info>=(3,0,0):
                    exec('print("", end=" ")')
                else:
                    exec('print "", ')
                GeometryInfo(subgeom,Prefix)
    else:
        print(line)
    return

File='C:/Data/LKA_adm0.shp'
DataSource=ogr.Open(File,update=0)
if DataSource is None:
    print 'Unable to open file'
    sys.exit(1)
print("Data source: %s" %(DataSource.GetName()))
driver=DataSource.GetDriver()
print("Driver: %s" %(driver.GetName()))
for iLayer in range(DataSource.GetLayerCount()):
    layer=DataSource.GetLayer(iLayer)
    layer_defn=layer.GetLayerDefn()
```

```
    print("")
    print("Layer name: %s" %layer_defn.GetName())
    print("Geometry: %s" %ogr.GeometryTypeToName(layer_defn.GetGeomType()))
    print("Feature count: %d" %layer.GetFeatureCount(force=1))
    oExt=layer.GetExtent(True)
    if oExt is not None:
        print("Extent: (%f, %f) - (%f, %f)" %(oExt[0],oExt[1],oExt[2],oExt[3]))
    if layer.GetSpatialRef() is None:
        srs_WKT="(unknown)"
    else:
        srs_WKT=layer.GetSpatialRef().ExportToPrettyWkt()
    print("Layer SRS WKT:\n%s" %srs_WKT)
    if len(layer.GetFIDColumn())>0:
        print("FID Column: %s" %layer.GetFIDColumn())
    if len(layer.GetGeometryColumn())>0:
        print("Geometry Column: %s" %layer.GetGeometryColumn())
    for iAttr in range(layer_defn.GetFieldCount()):
        Field=layer_defn.GetFieldDefn(iAttr)
        print("%s: %s (%d.%d)" % ( \
                    Field.GetNameRef(), \
                    Field.GetFieldTypeName(Field.GetType()), \
                    Field.GetWidth(), \
                    Field.GetPrecision()))
    Feature=layer.GetNextFeature()
    while Feature is not None:
        FeatureInfo(Feature)
        Feature=layer.GetNextFeature()
m = Basemap(projection='tmerc',lon_0=80.7,lat_0=7.9, \
        k_0=0.9996,rsphere=(6378137.00,6356752.314245179), \
        width=320000,height=480000,resolution='i')
(filepath,filename)=os.path.split(File)
(shortname,extension)=os.path.splitext(filename)
shapefile=filepath+'/'+shortname
s=m.readshapefile(shapefile,name='admin',drawbounds=True,linewidth=0.5, \
            color='k',antialiased=1)
plt.title("Sri Lanka",fontsize=15)
DataSource=m=None
```

CONTOUR GENERATION

```
#!/usr/bin/env python
import osgeo.gdal as gdal
import osgeo.ogr as ogr
import osgeo.osr as osr
import matplotlib.pyplot as plt
import sys,os,time
infile=r'C:\Data\srtm_45_07.tif'
dataset=gdal.Open(infile)
if dataset is None:
   print 'Unable to open file'
   sys.exit(1)
band=dataset.GetRasterBand(1)
spatial_ref=osr.SpatialReference()
spatial_ref.ImportFromWkt(dataset.GetProjection())
NoDataVal=band.GetNoDataValue()
geotrans=dataset.GetGeoTransform()
outfile=raw_input('Enter output contour shapefile name (including path): ')
field_name=raw_input('Enter field name for elevation data in contour shapefile: ')
ogr_ds=ogr.GetDriverByName('ESRI Shapefile').CreateDataSource(outfile)
(filepath,filename)=os.path.split(outfile)
(shortname,extension)=os.path.splitext(filename)
ogr_lyr=ogr_ds.CreateLayer(str(shortname),spatial_ref)
field_defn=ogr.FieldDefn('ID',ogr.OFTInteger)
ogr_lyr.CreateField(field_defn)
field_defn=ogr.FieldDefn(str(field_name),ogr.OFTReal)
ogr_lyr.CreateField(field_defn)
condition=raw_input('Enter 1 for specific contour generation; else 2 for \
               generating contours based on contour interval: ')
if int(condition) is 1:
   FixedLevels=raw_input('Enter space separated elevation data: ')
   t1=time.clock()
   FixedLevels=[float(x) for x in FixedLevels.split()]
   gdal.ContourGenerate(srcBand=band,contourInterval=0,contourBase=0, \
   fixedLevelCount=FixedLevels,useNoData=0,noDataValue=float(NoDataVal), \
   dstLayer=ogr_lyr,idField=0,elevField=1)
   t2=time.clock()
elif int(condition) is 2:
   contour_intvl=raw_input('Enter contour interval: ')
   BaseContourValue=raw_input('Enter elevation of base contour: ')
   t1=time.clock()
   gdal.ContourGenerate(band,float(contour_intvl),float(BaseContourValue), \
                  [],0,float(NoDataVal),ogr_lyr,0,1)
```

```
    t2=time.clock()
print 'Contours were generated in %f seconds'%(t2-t1)
min=band.GetMinimum()
max=band.GetMaximum()
if min is None or max is None:
    MinMax=band.ComputeRasterMinMax(False)
    min=MinMax[0]
    max=MinMax[1]
BandData=band.ReadAsArray()
fig1=plt.figure(figsize=(9,7),facecolor='white')
fig1.canvas.set_window_title("SRTM data plot")
ax1=fig1.add_subplot(1,1,1)
cax=ax1.imshow(X=BandData,vmin=min,vmax=max,alpha=1,cmap=plt.cm.terrain)
cbar=fig1.colorbar(cax,ticks=[min,max],orientation='vertical')
cbar.ax.set_yticklabels([str(min),str(max)])
ax1.set_xlabel('Longitude (degree)')
ax1.set_ylabel('Latitude (degree)')
ax1.set_title("SRTM data plot")
ax1.set_xticks((0,1200,2400,3600,4800,dataset.RasterXSize))
x=int(geotrans[0])
ax1.set_xticklabels((x,x+1,x+2,x+3,x+4,x+5))
ax1.set_yticks((0,1200,2400,3600,4800,dataset.RasterYSize))
y=int(geotrans[3])
ax1.set_yticklabels((y,y-1,y-2,y-3,y-4,y-5))
ax1.grid(True)
pix=dataset.RasterXSize*dataset.RasterYSize
fig2=plt.figure(figsize=(9,7),facecolor='white')
fig2.canvas.set_window_title("Histogram of SRTM data")
ax2=fig2.add_subplot(111)
BandData=BandData.reshape(pix,)
ax2.hist(x=BandData,bins=100,range=(min,max),facecolor='green',alpha=1.0, \
        histtype='bar',align='mid',orientation='vertical')
ax2.set_xlabel('Elevation (meter)')
ax2.set_ylabel('Frequency')
ax2.set_title('Histogram of SRTM data')
ax2.set_xlim(min,max)
ax2.grid(True)
dataset=fig=None
ogr_ds.Destroy()
```

RASTER CLIPPING

```
#!/usr/bin/env python
import osgeo.gdal as gdal
import osgeo.gdal_array as gdal_array
import osgeo.gdalconst as gdalconst
import osgeo.ogr as ogr
import osgeo.osr as osr
import math,numpy
import matplotlib.pyplot as plt
gdal.AllRegister()
inRaster='C:/Data/SatImage.tif'
inVector='C:/Data/county.shp'
outMask='C:/Results/outMask.tif'
outRaster='C:/Results/outRaster.tif'
driver=gdal.GetDriverByName('GTiff')
inDataset=gdal.Open(inRaster,gdalconst.GA_ReadOnly)
inRasterData=gdal_array.DatasetReadAsArray(inDataset)
in_r_srs=osr.SpatialReference()
in_r_srs.ImportFromWkt(inDataset.GetProjection())
in_DataType=inDataset.GetRasterBand(1).DataType
in_r_geo=inDataset.GetGeoTransform()
in_r_col=inDataset.RasterXSize
in_r_row=inDataset.RasterYSize
in_r_band=inDataset.RasterCount
in_r_ulx,in_r_uly=in_r_geo[0],in_r_geo[3]
in_r_pixelX,in_r_pixelY=in_r_geo[1],-in_r_geo[5]
inShapefile=ogr.GetDriverByName('ESRI Shapefile').Open(inVector,0)
in_v_minX,in_v_maxX,in_v_minY,in_v_maxY=inShapefile.GetLayer(0).GetExtent()
offsetX=math.floor((in_v_minX-in_r_ulx)/in_r_pixelX)
out_r_ulx=in_r_ulx+in_r_pixelX*offsetX
offsetY=math.floor((in_r_uly-in_v_maxY)/in_r_pixelY)
out_r_uly=in_r_uly-in_r_pixelY*offsetY
out_r_geo=(out_r_ulx,in_r_pixelX,0,out_r_uly,0,-in_r_pixelY)
out_r_col=int(math.ceil((in_v_maxX-out_r_ulx)/in_r_pixelX))
out_r_row=int(math.ceil((out_r_uly-in_v_minY)/in_r_pixelY))
mask=gdal.GetDriverByName('MEM').Create('',out_r_col,out_r_row,1, \
                    gdalconst.GDT_Byte)
mask.SetGeoTransform(out_r_geo)
mask.SetProjection(in_r_srs.ExportToWkt())
mask_band=mask.GetRasterBand(1)
mask_band.Fill(0)
gdal.RasterizeLayer(dataset=mask,bands=[1],layer=inShapefile.GetLayer(0), \
          burn_values=[255],options=["ALL_TOUCHED=TRUE"])
```

```
mask_data=mask_band.ReadAsArray(0,0,out_r_col,out_r_row)
gdal_array.SaveArray(mask_data,outMask,format="GTiff",prototype=mask)
inRasterData_trim=inRasterData[:,offsetY:(offsetY+out_r_row), \
                        offsetX:(offsetX+out_r_col)]
outDataset=gdal.GetDriverByName('MEM').Create('',out_r_col,out_r_row, \
                             in_r_band,in_DataType)
outDataset.SetGeoTransform(out_r_geo)
outDataset.SetProjection(in_r_srs.ExportToWkt())
for iBand in range(outDataset.RasterCount):
   in_band=inDataset.GetRasterBand(iBand+1)
   NoDataVal=in_band.GetNoDataValue()
   outDataset.GetRasterBand(iBand+1).SetNoDataValue(NoDataVal)
outRasterData=gdal_array.DatasetReadAsArray(outDataset)
opt=raw_input('''Enter 1 for raster clipping to feature extent,
           2 for raster clipping to feature shape: ''')
if int(opt)==1:
   outRasterData=numpy.copy(inRasterData_trim)
elif int(opt)==2:
   condlist=[mask_data==0]
   choicelist=[int(NoDataVal)]
   outRasterData=numpy.copy(inRasterData_trim)
   outRasterData=numpy.select(condlist,choicelist,inRasterData_trim)
gdal_array.SaveArray(outRasterData,outRaster,format="GTiff",prototype=outDataset)
inRasterData=numpy.rollaxis(inRasterData,0,3)
fig1=plt.figure(figsize=(9,7),facecolor='white')
fig1.canvas.set_window_title("Original data")
ax1=fig1.add_subplot(1,1,1)
cax=ax1.imshow(X=inRasterData)
ax1.axis("off")
outRasterData=numpy.rollaxis(outRasterData,0,3)
fig2=plt.figure(figsize=(9,7),facecolor='white')
fig2.canvas.set_window_title("Clipped data")
ax2=fig2.add_subplot(1,1,1)
cax=ax2.imshow(X=outRasterData)
ax2.axis("off")
inDataset=outDataset=mask=None
inShapefile.Destroy()
```

LIBLAS INFORMATION

```
#!/usr/bin/env python
import liblas
print 'libLAS version: ',liblas.get_version()
print 'Python version: ',liblas.version
print 'Installation path: ',liblas.__path__
print 'GDAL enabled/linked: ',liblas.HAVE_GDAL
print 'LibGeoTIFF enabled/linked: ',liblas.HAVE_LIBGEOTIFF
```

READING LAS HEADER

```
#!/usr/bin/env python
import liblas
in_file='C:\\Data\\Barrow_SeaIce_May7_2008.laz'
f=liblas.file.File(filename=in_file,mode='r',in_srs=None,out_srs=None)
hdr=f.header
print 'File signature: ',hdr.file_signature
print 'File source ID: ',hdr.filesource_id
print 'Project ID: ',hdr.project_id
print 'LAS file version: ',hdr.version
print 'Version major: ',hdr.major_version
print 'Version minor: ',hdr.minor_version
print 'System identifier: ',hdr.system_id
print 'Generating software: ',hdr.software_id
print 'Header size(in bytes): ',hdr.header_length
print 'Offset to point data (in bytes): ',hdr.data_offset
print 'Number of Variable Length Records: ',hdr.records_count
print 'Point data format ID: ',hdr.dataformat_id
print 'Number of point records: ',hdr.point_records_count
print 'Number of points by return: ',hdr.return_count
print 'Scale factor: ',hdr.scale
print 'Offset: ',hdr.offset
print 'Min: ',hdr.min
print 'Max: ',hdr.max
print 'Compressed file: ',hdr.compressed
f.close()
```

READING LAS POINTS

```
#!/usr/bin/env python
import liblas
in_file='C:\\Data\\Barrow_SeaIce_May7_2008.laz'
f=liblas.file.File(in_file,mode='r')
hdr=f.header
for i in range(f.__len__()):
   print 'Record ',i+1
   pnt=f.read(i)
   print 'X,Y,Z: ',pnt.x,pnt.y,pnt.z
   print 'Intensity: ',pnt.intensity
   print 'Return number: ',pnt.return_number
   print 'Number of returns: ',pnt.number_of_returns
   print 'Scan direction flag: ',pnt.scan_direction
   print 'Edge of flight line: ',pnt.flightline_edge
   print 'Classification: ',pnt.classification
   print 'Scan angle rank: ',pnt.scan_angle
   print 'User data: ',pnt.user_data
   print 'Point source ID: ',pnt.point_source_id
   if hdr.data_format_id==1 or hdr.data_format_id==3:
      print 'GPS time: ',pnt.time
   if hdr.data_format_id==2 or hdr.data_format_id==3:
      c=pnt.color
      print 'Red,Green,Blue colors: ',c.red,c.green,c.blue
f.close()
```

WRITING LAS FILE

```
#!/usr/bin/env python
import liblas
in_filename='C:\\Data\\Barrow_SeaIce_May7_2008.laz'
in_file=liblas.file.File(in_filename,mode='r')
in_hdr=in_file.header
out_filename='C:\\Results\\Output.las'
out_hdr=liblas.header.Header()
out_file=liblas.file.File(out_filename,mode='w',header=out_hdr)
out_hdr.major=in_hdr.major
out_hdr.minor=in_hdr.minor
for p in in_file:
   pnt=liblas.point.Point()
   if p.scan_angle==0:
      pnt.x=p.x
      pnt.y=p.y
      pnt.z=p.z
      out_file.write(pnt)
in_file.close()
out_file.close()
```

www.ingramcontent.com/pod-product-compliance
Lightning Source LLC
Chambersburg PA
CBHW051017180526
45172CB00002B/388

9781479316472